Global Diversities

Over the past decade, the concept of 'diversity' has gained a leading place in academic thought, business practice, politics and public policy across the world. However, local conditions and meanings of 'diversity' are highly dissimilar and changing. For these reasons, deeper and more comparative understandings of pertinent concepts, processes and phenomena are in great demand. This series will examine multiple forms and configurations of diversity, how these have been conceived, imagined, and represented, how they have been or could be regulated or governed, how different processes of inter-ethnic or inter-religious encounter unfold, how conflicts arise and how political solutions are negotiated and practiced, and what truly convivial societies might actually look like. By comparatively examining a range of conditions, processes and cases revealing the contemporary meanings and dynamics of 'diversity', this series will be a key resource for students and professional social scientists. It will represent a landmark within a field that has become, and will continue to be, one of the foremost topics of global concern throughout the twenty-first century. Reflecting this multi-disciplinary field, the series will include works from Anthropology, Political Science, Sociology, Law, Geography and Religious Studies. While drawing on an international field of scholarship, the series will include works by current and former staff members, by visiting fellows and from events of the Max Planck Institute for the Study of Religious and Ethnic Diversity. Relevant manuscripts submitted from outside the Max Planck Institute network will also be considered.

Sakura Yamamura

Spatial Diversity in the Global City

Transnational Tokyo

palgrave
macmillan

Sakura Yamamura
Department of Geography, RWTH Aachen University
Aachen, Nordrhein-Westfalen, Germany

ISSN 2662-2580 ISSN 2662-2599 (electronic)
Global Diversities
ISBN 978-3-031-64724-6 ISBN 978-3-031-64725-3 (eBook)
https://doi.org/10.1007/978-3-031-64725-3

Cover illustration: Kawee Wateesatogkij / Alamy Stock Photo

This Palgrave Macmillan imprint is published by the registered company Springer Nature Switzerland AG.
The registered company address is: Gewerbestrasse 11, 6330 Cham, Switzerland

To Frau Dr. Philipp
Deutsche Schule New York
(now GISNY)

Preface

The global COVID-19 pandemic hit the world some time after the original manuscript was finalized. Especially the lockdown phase and the shocking closing of borders may cast doubt on the timeliness and relevance of this piece on the growing mobility of transnational migrants and their conviviality in urban spaces. However, the pandemic actually has exhibited how much issues of the global mobility or the restraints thereof as well as the challenges of growing diversification in cities and elsewhere are more than topical and important to be explored.

At first glance, what comes to mind when confronted with the COVID-19 crisis is the shock and the accompanying anxious feeling of vulnerability of being in a lockdown, confined in the free mobility and restricted in the socialization. The drastic constraint on mobility and the physical as well as social distancing we were obliged to appear paradoxical to any debates on conviviality in cities. These aspects might appear to have halted the discourses on the hypermobility of people for the time being. In a brief moment, indeed, the hype and even partly glorification of the emergence of a new group of global citizens, characterized by their global mobility and their societal influence spanning the globe seemed to vanish. It is with a bitter note, maybe in some cases even with a slight spitefulness, that we seemed to be locked down globally and affected by immobility universally.

However, what is more striking than this initial reaction and impression was, in fact, that the pandemic has shown the power of migration policies and law enforcements of national government more than ever, and there was a twist as to the global mobility of individuals now popping up as global nomads working from anywhere in the world since the pandemic has eased corporate mobile work policies. The pandemic actually has crystallized the privilege of mobility for the still traveling global elites and superrich on the one hand (after a brief interruption and relative inconvenience), and the confinement of the less privileged, leaving them in desperation and total impotence and immobility.

Moreover, as to the seemingly ever-shrinking global community, that had long been engraved into our brains with mantra-like repetition of slogans such as "Think global(ly), act local(ly)," the ever-increasing global connectedness with social media and the buzz regarding the urbanization of human society—with the United Nations proclaiming the era of urban population in 2014 when more than half of the world population lived in cities—the pandemic has also shown its darkest side of these hypes. Coinciding with Black Lives Matter movements, the pandemic has spawn subliminally extant anti-Asian racism to the surface in many Western countries. At the same time, the pandemic caused previously unthinkable forces of divergent stances from the alt-right, spiritual groupings, left-wing anti-vaxxers and other conspiracists to demonstrate together and show discontent in a manner that had never been seen before. The global society and the growing urban diversity appeared to collapse in dire straits.

It is against such background that the global question on the mobility and the conviviality of diversifying urban societies arises again. It has actually never been more topical and relevant before, questioning the mechanism of freedom and constraints to mobility. History shows, the memory of society tends to be short-lived, and already after three years, we see that the global society is catching up with the lifestyle as before; with indeed focusing now on the next crisis of the Russia-Ukraine war, the energy crisis, yet another Middle East conflict, there is a fading sense of the restrictions and shock that the global pandemic had brought us. Borders have started to be re-opened, and the global mobility of tourists, professionals, students and all other migrants is picking up again. Borders are now being closed again, but not due to the pandemic; it is—back to the state before—due to refugee and migration policies at large.

Presumably, it is by no time that the global hype on the dominating cities will be glaring on the global fora of the economy and politics, with regular reports and rankings of global consultancies and the like on the Global Cities Index (A.T. Kearney), Global City Competitiveness Index (The Economist Group), Global cities Survey (Knight Frank LLP with Citi Private Bank) or the Global Power City Index (The Mori Memorial Foundation). These cities will be illustrated as the place to be, where flourishing diversity is lived by its fullest. Hyped again will be, or partly are already, the hypermobiles, global expats or the transnational capitalist class. Hailed as the personification of the global change in society and the new wave of hybrids, global nomads or even "mongrels," in creative non-fiction and journalistic anecdotal accounts of Global Me's and Global Souls, this category of transnational mobiles appears once again to be the key to understanding the socio-spatial dynamics of the recent world.

In all these hyping for the global mobility and connectedness, yet also hailing the peculiarity of the local and the urban, transnational corporations appear to be especially those which are internalizing but also amplifying this idea. Advertisements for transnational financial institutions, such as ING, Citibank or HSBC—which claims to be "the world's local bank"—bombard the mobile traveler with their creative interpretation of the (business) world at major international airports, suggesting a vibrant and dynamic world being increasingly connected ("Local demand will shape global supply," "There's a new world emerging") and calling for a global solidarity and responsibility, but also for the recognition and acceptance of the diversity of cultures and society ("Different values make the world a richer place"). These ads are clearly targeting the very specific group of customers who travel frequently and have a higher probability of noticing them, who are, in their words, forming "a high value, influential and cosmopolitan audience" (JCDecaux 2013). Flattering them with pictures in the gangways to the aircraft illustrating the impact that they could make to the world but also the very privilege these travelers have ("Unlock a world of unforgettable experiences"), the ads of these global players go so far as to even suggest the hip way of experiencing a city

("Tokyo, here I come" in collaboration with the global backpacker's guide Lonely Planet).

But, where do they—these hypermobile transnational professionals and migrants—go to and come from? Is this locational advice that such transnational financial corporations and other global players give and the spaces they convey the real "local" Tokyo? Where are the spaces located, in which such hypermobile transnational professionals live and work? Is there a specific space produced and even reproduced by these mobile transnational professionals and their transnational corporation employers?

That there is a social diversification occurring around the world with millions of individuals moving and that this also inevitably results in spatial diversification changing the "original" urban landscape seem obvious, but little is really known about the process of such socio-spatial diversification in urban spaces. Diversification, if not superdiversification, seems to be everywhere, but indications of where it exactly finds its spatial expression usually do not go further than generic labels of the city being a global city, multicultural city or superdiverse city. Where, however, are these spatial diversifications caused by the much-observed transnational migration in particular, and where are the production and reproduction of transnational spaces within the global arena of cities located?

This work delves deeper into the question for the *where* and *how* of the transnational space in the context of the diversifying global cities emerging and spreading in the world by bringing together following three aspects: (a) the transnational professionals as part of the growing transnational migrant population, (b) the transnational corporations in the global economy as the framework in which these transmigrants are embedded, and (c) the network of global cities where each city has its own local spatialities. By doing so it bridges the idea of merely looking at migration *to* and *in* cities to a notion of migrants *and* cities, and by this bringing the geography of the much-discussed, and well-hyped, concept of superdiversity on the geographical agenda.

As a final note to the reader: This monograph is originally based on a dissertation. While the main empirical material still stems from the fieldwork in that context, the monograph has been complemented by further fieldworks conducted in the following years, and it also contains novel theoretical discussions and thoughts evolving in the course of the academic path. The monograph is conceptualized in a maybe scientifically unusual style in that each chapter has its own narrative and its own style and can be read independently as stand-alone. While some chapters, such as Chaps. 2, 3 and later 9 are closer to the academic tradition of referenced literary and theoretical discussion, other parts especially in the middle, like Chaps. 4–6, are empirical and more vivid accounts of different spaces within Tokyo. For those who are more intrigued by the phenomenon of global changes of urban diversities and interested in the gist of the scholarly debate and the novel thoughts on spatial superdiversity, Chaps. 1, 8 and 9 would be recommendable. Thus, the reader is welcome to read through the monograph from the beginning to the end, but also invited for quick reads to jump and skip through focusing on the one or other topic by liking.

Aachen, Germany Sakura Yamamura

Acknowledgment

This manuscript has journeyed far, beginning as a dissertation and evolving throughout my postdoctoral period into its current form. Two vital lessons have been imparted along this academic path: first, that a work is never truly complete or final—it must simply be released at some point and allowed to stand on its own as it is. The second lesson is that it is impossible to thrive without the support of others. I would like to take this opportunity to express my deep gratitude to everyone who has accompanied both me and this work.

My first and foremost thanks are reserved for my PhD supervisors, Christof Parnreiter and Steven Vertovec. My acknowledgment of them here is not merely a formality; it is a sincere expression of appreciation. Regarding Herr Parnreiter, I would like to express my immeasurable gratitude for the freedom and support I have received over the years. Similarly, I owe a great debt to Steve for his mentorship and the invaluable advice he has given me, which helped me navigate the complexities of the academic jungle. Thank you for taking me under your wings.

For any researcher indispensable and invaluable: all the interviewees. I especially greatly appreciate my gatekeepers Pedro Furumoto, Rui Fukuda, Yamamoto-kun and Mizuta-kun, but also all 45 transnational professionals, as well as the experts and further interviewees that followed, for their time and interest. I hope I could do justice for your efforts with the present work. I would also like to express my gratitude to

Mesdames de Yamanote for their precious time and highly intriguing lunch time.

My parents provided the greatest support in accessing these resources and shaping my life; words cannot fully express my gratitude. Thank you for everything. I also want to thank my siblings and dear friends for their enduring support. Special thanks go to Paul Lassalle, my sparring partner, whose friendship and collegiality have been invaluable throughout this challenging journey.

Last but not least, I dedicate this work to my one and only geography teacher, Frau Dr. Philipp from my time at the German School New York. Her awe-inspiring insightfulness and far-sightedness imbued me with the interdisciplinary and holistic geographical perspective. Had it not been for her competence and passion for geography, I would have never found my own passion for this discipline. With my deepest respect and gratitude, I dedicate this work to the person who set me on the journey into geography already very long time ago.

Praise for *Spatial Diversity in the Global City*

"Set in the global city of Tokyo, this book zooms in on the complexities and paradoxes in the way transnational professionals engage in space-making. By carefully differentiating between social spaces – from the comfort zones of the foreigner ghetto to the contact zones in local society – it yields fascinating insights on the spatial and temporal aspects of urban diversification."

—Brenda Yeoh, *Raffles Professor in Social Sciences, National University of Singapore*

"For a city of its size and global importance, Tokyo is strangely under-researched. All the more welcome therefore is this volume that sits on the interface of urban and migration studies. Sakura Yamamura casts expert light on the lives of transnational professionals in Tokyo, how they interact with the city and how the city shapes them, offering on the way telling insights on the nature of diversification and convergence in global cities."

—Paul Waley, *Honorary Research Fellow, University of Leeds, UK*

Contents

List of Figures

1

Introducing the Debate

> Within the migration literature there are many studies of migration *to* cities and the life of migrants *in* cities but very little about the relationship of migrants *and* cities. However, in most of this scholarship cities figure merely as containers providing spaces in which migrants settle and make living. Despite the vast migration scholarship on what are labeled ethnic, transnational, or diasporic communities in specific cities, we still know very little about how migrants actively contribute to the restructuring and repositioning of either their cities of settlement or those to which they are transnationally connected. (Çağlar & Glick-Schiller, 2011: 2)

Debates on transnationalism which burgeoned as early as in the late 1980s in social scientific migration research, if not even earlier in economic geography and international business studies in the context of corporate transnationalization, gained momentum in the recent decades, a time marked by increased mobility of goods, information and especially people to an unprecedented global extent. It was also described as the emergence of the "space of flows" (Castles & Miller, 1998). With the transnationalism agenda moving forward and even disemboguing into the newer stream of diversity studies, what has been whatsoever passed

S. Yamamura, *Spatial Diversity in the Global City*, Global Diversities,
https://doi.org/10.1007/978-3-031-64725-3_1

over in these debates surrounding urban superdiversity is the spatial aspect of the transnational migrations particularly in the context of the urban space. While debates on transnational urbanism have synthesized the urban with the transnational migration (Smith, 2000), the actual dynamics of diversifications in and of the urban space have been overlooked. Cities, however, are where the superdiversity of the migrant population or the migrant-led diversity (Vertovec, 2007), including that of transnational migrants, concentrates and where inevitably socio-spatial diversification must also be observable. Cities should be regarded neither in the sense of vessels accommodating migrants on their journey nor as preconfigured destinations in which migrants settle. They should be understood as a scaled unit within a larger structure of global economy and global society, but also as a locational constitution of society formed by the transnational migrants themselves. In the terminology of geography, the city should neither be exclusively understood as a territorial space providing migrants structures and the constellation in which they adapt to (city *as* space) nor solely in relational terms as constructs within a larger spatial context to which migrants move back and forth (city *in* space) (Lichtenberger, 1986).

The key to understanding the spatial diversification in the context of increasing transnational migration lies in the bridging of these spatial concepts by focusing on the idea of, as Çağlar and Glick Schiller point out, migrants *and* cities. By taking the spatial conceptual approach of studying the transnational space, this monograph opens new floor to understanding the complex interrelation between the migrants *to*, *in* and *and* cities. The transnational space is understood as a spatial phenomenon both spanning over the globe but also as being locatable within the global network and even more so within the city. In this particular case, the study of space of migrants *and* cities, more precisely the transnational space of transnational migrants and global cities, contributes to the understanding and theorizing of the spatial diversification emerging in the times of increasing mobility.

Global cities in this context provide a particularly interesting field of research into spatial diversification. These cities, understood as command and control centers of global economy (Sassen, 2002), are under specific dynamics through the constant flow of capital, goods and people (Castles

& Miller, 1998). These flows are evidently occurring globally, but this supposedly universal trend is limited primarily to the network of global cities, whereas the intensity of similar developments can be assumed to be higher with the interconnectedness of the particular global city in the network (Taylor & Derudder, 2015). At the same time, the currents of social dynamics are especially strong in these cities. A high diversity of migrants flow in and out and also stay at different time spans in such cities, accommodating an increasingly different profile of migrants as pointed out by the term of superdiversity, encompassing differences in legal status, migration channel, duration, skills, but also personal attributes (Vertovec, 2007). Global cities are evidently the ideal place to research spatial diversification processes in the context of transnational migrations.

This monograph focuses on uncovering such socio-spatial dynamics by taking an actor-based empirical perspective on a group of transnational migrants who have been neglected as migrants themselves: the transnational migrant and corporate professionals. It furthermore scrutinizes this particular group of migrants in a global city which have also been overlooked in scholarship, the global city-region of Tokyo. Apart from their role as agents of transnational corporations acting in the global economy, thus, as actors of "transnationalism from above," transnational professionals did not find much attention in the actual migration research.[1] In fact, they have been discussed primarily in the expatriate research and transnational business communities, but not as part of the debates surrounding transnational urbanism. In other words, this monograph sheds new light on the making of the transnational space which spans *in* and *over* the global city network including cities such as Tokyo.

In accordance with the critique of Çağlar and Glick Schiller (2011, see above), this book contributes to extending the understanding of the spatial diversification of transnational urban space from different levels. From the macro- and meso-level perspective, it is about flows of

[1] The terms "transnational professionals," "transnational corporate professionals," "transnational migrant professionals" and similar will be used interchangeably throughout the book for the purpose of variety for the reader; what is meant by them are highly skilled and highly mobile professionals who work for transnational corporations and who should be regarded part of transnational migrants.

corporate and highly skilled migrants into specific countries and cities. On the micro-level, the focus on the everyday living of migrants in such global cities uncovering their socio-spatial patterns. It is an in-depth exploration of the transnational spaces produced and reproduced by these transmigrants who are located both within and beyond the global city of Tokyo and within the "transnationally connected" (ibid.) network of global cities. Whereas the *relational approach* to cities has been focusing on the overall structures of city networks and the causes of the creation of such networks from a top-down view, parallel to the "transnationalism from above," thus focusing on the global economy with its global players in the form of transnational corporations or governance institutions, the more *territorial approach* to cities has been focusing on the constellation of such cities in which people including migrants unfolded social and spatial dynamics within the given urban constellation, parallel to the "transnationalism from below."

By acknowledging both, the structural impacts of the dynamics of global economy on the production and emergence of global cities, as well as the specific constellation of these global cities in being not only created by ghost hands of global economy but by specific actors in them, the monograph contributes to addressing the gaps in the debate on spatial diversification in cities *and* migration.

Concerning these specific actors, transnational professionals, especially of transnational corporations of the financial industry, take a special role. On the one hand, transnational professionals of firms directly and primarily impact the emergence of global cities, that is, particularly those of the FIRE sector (finance, insurance and real estate). They take the role of being the global city makers; thus, their economic decisions impact the spatial constellation of the global city network but also within the global cities. On the other hand, they are also simultaneously individuals with own socio-spatial patterns as any other migrant, worker or resident in the city, by that, impacting and creating the transnational space through their everyday practices. It is not in their role as corporate professionals, but being migrant individuals in society with one of highest capitals, in particular economic and symbolic capital, they consciously or unconsciously, substantially influence societal dynamics.

Despite this prominent role held by this specific group of transnational migrants, they are so far under-researched. As briefly mentioned above, usually transnational professionals are regarded and also self-perceive themselves as "expats" (expatriates) and not migrants. In academia, too, transnational professionals have been mostly researched in the context of transnational corporate human resource strategies or as part of their function in transnational business and corporate networks. However, not only in legal and migration policy terms, transnational professionals actually are the epitome of transnational migration. With their high mobility and repetitive mid-term relocations,[2] they do not only migrate from one country to another. They even migrate to multiple countries and keep social ties not only in the classic sense of the transnational migration between the country of origin and destination, but further to multiple countries and societies.

Finally, in this context, the dynamics of these destinations for the flows of transnational migrants need to be reconsidered. As much as the global cities remain in the network of such peer cities once they develop interrelational economic ties, there are also dynamics within the network, too. While cities as New York, London and Tokyo have been broadly accepted as the top global cities as early as the 1990s, the global network has expanded over the decades with emerging cities especially in the Asian region. Singapore and Hong Kong have come into focus of economic investments but also of academic research, drawing away interest in Tokyo. Questions have been arising on the diverging developments of these Asian global cities compared to the Anglo-American and European cities on one side. On the other, these cities are in contrast to Tokyo, so to say, "by nature" diverse as postcolonial and bi- or even multilingual cities, differing from other non-English based global cities and probably further future global cities emerging in other parts of the world.

There are several intentions behind the locational choice of Tokyo for the study of spatial diversification in global cities. First, Tokyo has been

[2] International migration is by international definition by the UN relocations between countries of more than one year duration, whereas those below are administratively and legally conceived as short-term and those exceeding five years are seen normally as either long-term or even permanent migration as in most OECD countries permanent residence permits can be obtained from five years' legal sojourn.

identified and established as one of the top global cities right in 1990 by Sassen and widely accepted as such, yet empirical research—probably due to also linguistic barriers—remained undone in contrast to the abundance of research on the other two global cities, that is, New York and London. Second, among all the nowadays well-established global cities of Asia, with Singapore, Hong Kong but also Shanghai, the focus of not only economy but also academia turned away from Tokyo. While this parallels the economic downturn and subsequent stagnation of Japanese economy, it leaves the question of what happens with global cities in such new currents and dynamics of global economy and mobility. Thirdly, the unique case of Tokyo as being a non-English-speaking and non-postcolonial city could actually become an interesting case for the future of cities as increasingly cities not following the footsteps of the migrant populated global cities might be well emerging from other parts of the world, be it Arab cities or Southeast Asian cities. This means that Tokyo as the currently unique case might become a prototype or at least a valuable case for non-English-speaking non-postcolonial cities in future.

This monograph is embedded in the larger framework of questioning the nature of spatial diversification induced by the increasing transnational migrations to cities worldwide. Within this broader context, it follows the research question on where the transnational space(s) of the specific group of transnational migrants can be found. Further, by empirically exploring the socio-spatial patterns of the transnational professionals, it also discusses how these transnational spaces are constituted in the particular case of global city Tokyo as well as in the overall global network, which then closes the question back to the spatial diversifications in global cities.

As the Preface has pointed out, we are living in unstable and unpredictable times, where focus of society changes quickly, and we seem to be moving from one crisis to another. It might even seem that in-depth research into a city and its mobile residents might become quickly obsolete. However, socio-spatial dynamics in urban spaces and in the network of cities remain crucial and relevant as the era of global mobility and connectedness—irrespective of momentary immobility and confinements as during the pandemic—cannot and will not be turned back any time soon. It is of utmost importance that we do scrutinize dynamics of the socio-spatial diversifications in cities, the dynamics between migrants *and* cities, as these are the key to a more peaceful conviviality for our future global society.

After setting the overall scene of the monograph by introducing the recent theories and debates on transnationalism and spatial diversification and the adjacent debates on global cities in Chap. 1, a concise literature review on Tokyo regarding the global city issue is provided in Chap. 2. Chapter 3 discusses the spatial theoretical concept in which the empirical study is embedded, delivering the scientific grounding of the study. As inherent part of the conceptual embedding, the methodological approach is also laid out there and the research design as well as the conclusive research process explicated. On these bases, the following Chaps. 4–6 present the empirical findings of the different socio-spatial patterns within the local level of global city Tokyo. The discursive chapters following these empirical findings conceptually abstract the findings into debates on the spatial extension of transnational space to the regional level (Chap. 7) and also by complementing with further empirics bringing the debate on socio-spatial diversification into a global level (Chap. 8). The conclusion reviews the main findings and their conceptual implication presented in the preceding chapters, and finally identifies areas for further research. In reference to the Preface, as major societal changes occurred after the study through the COVID-19 pandemic, the monograph closes with a Coda which reflects the study results and contributions against the background of recent societal changes and contextualizes the urban socio-spatial dynamics in the ever-changing world society.

References

Çağlar, A., & Glick-Schiller, N. (2011). *Locating migration: Rescaling cities and migrants*. Cornell University Press.

Castles, S., & Miller, M. J. (1998). *The age of migration: International population movements in the modern world* (2nd ed., rev. and updated ed.). Macmillan.

Lichtenberger, E. (1986). Stadtgeographie–Perspektiven. *Geographische Rundschau, 38*(7), 388–394.

Sassen, S. (2002). *Global networks, linked cities*. Routledge.

Smith, M. P. (2000). *Transnational urbanism*. Wiley.
Taylor, P., & Derudder, B. (2015). *World city network: a global urban analysis*. Routledge.
Vertovec, S. (2007). Super-diversity and its implications. *Ethnic and Racial Studies, 30*(6), 1024–1054.

2

Setting the Scene: Transnationalism and Spatial Diversification in Global Cities

This chapter provides a concise overview of the debates surrounding the field of transnationalism especially in the context of urban space as relevant for this monograph. The aim is to set the overall scene to give a better understanding of the context in which the monograph's topic is embedded and to illustrate how different strands of research on transnationalism and urban studies are brought together to shed light on the socio-spatial diversification in global cities. Thus, it is not an attempt to deliver exhaustive literature reviews of each of the interconnected research strands but to present the gist of the debates.

The peculiarity of the relevant debates lies in the fact that two broader streams, each from different disciplinary backgrounds, basically flow around the topic of transnational professional migrants, creating a vacuum on this specific group of migrants who are in-between the so-called transnationalism "from above" and "from below." These different debates on transnationalism will be first discussed in this chapter. The second part then discusses the streams on the nexus on transnationalism and cities paralleling also the dichotomous disciplinary divide in a perspective from below (transnational urbanism) and from above (global cities). It is here where it will be clarified that the approach to search for the

S. Yamamura, *Spatial Diversity in the Global City*, Global Diversities,
https://doi.org/10.1007/978-3-031-64725-3_2

transnational spaces of transnational professionals is viable for studying the spatial diversification occurring in cities, in particular in global cities. While discussing the spatial diversification, a quick overview will be given on the research from the superdiversity perspective that have emerged more recently but rather detached from the global city research. In the context of these global cities, which are the city debates on the side of "transnationalism from above," the global city Tokyo will be introduced as an excursus. The chapter concludes with the illustration of how these diverging literature strands conceptually merge into the debate on transnational space of transnational migrant professionals as in this book.

Roughly sketching out the research direction on the nexus of transnationalism and cities, the two in each diverging storylines and foci can be summed up as follows (Fig. 2.1):

The first part of the chapter deals with the inner parts directly related to transnational professionals: On the one hand, as Fig. 2.1 depicts, international business studies and economic geography discuss the transnationality mainly in the context of transnational corporations and business

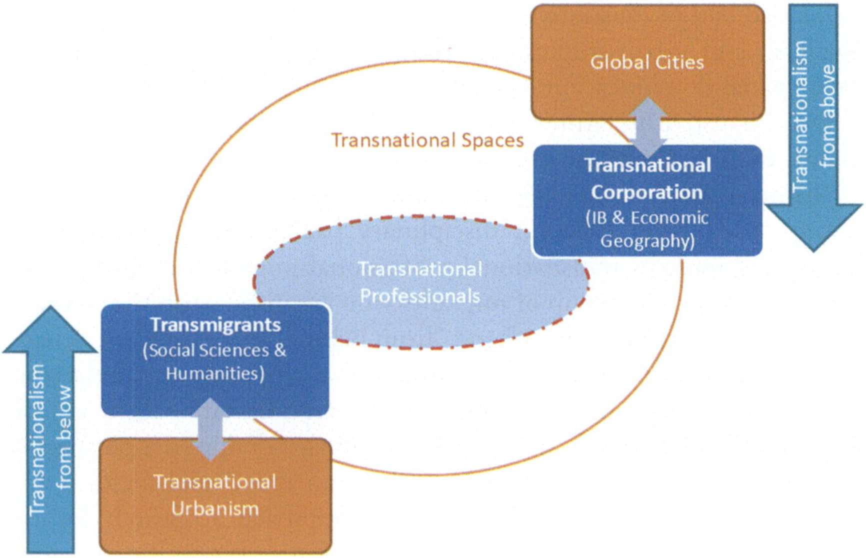

Fig. 2.1 Overview of research directions on the nexus of transnationalism and cities (author's own)

networks, which have been characterized mostly as being the "transnationalism from above." On the other hand, the transnationalism debates in social sciences and humanities, particularly migration studies, have been focusing, in part as a reaction to the previous movement "from above," on the "transmigrants from below," whereas a certain vacuum in-between these aboves and belows is created, that is, the field on transnational professionals.

The second part of the chapter explicates the strands surrounding the transnational spaces, depicted in the outer circle, whereas the disciplinary divide from the transnationalism research is also paralleled here. Whereas the transnational urbanism primarily orbits around transmigrants from below, the urban context of transnationalism from above is mainly located in the discussion of global cities. Though related to each other through the transnational migrants who socio-spatially interact and, thus, contribute to the urban space and its making, these two strands are rarely brought together. As this literature overview will show, the approach to discuss transnational professionals and their spaces in the context of transnational spaces particularly by recontextualizing them as a specific group of transnational migrants and in global cities is the key to closing the gap in the research. It also illustrates how this is also closely connected to the more recent and viable debates on spatial diversification in the context of superdiversity research.

The third part of this chapter then focuses on the debate on the specific case of the global city Tokyo, whereas its actual urban historical introduction is kept for later as frameworks for the empirical chapters, and the focus here is to clarify the relevance of the case in the research field. The overview on the research of the global city Tokyo will also clarify the empirical research gap existing in the current debate.

2.1 Transnationalism from Below and Above

The Transnational Turn in Migration Research

The "transnational turn" among migration scholars (King, 2012), in particular among anthropologists and sociologists, occurred in the early 1990s, especially with the theorization in the seminal work of Glick-Schiller et al. (1992). While migration research until then focused primarily on international and internal migrations (cf. Cohen, 1996; Massey et al., 2005), transnationalism researchers broke a new path within migration research (cf. Brettell & Hollifield, 2000; King, 2012), by pointing out the societal transformations caused by the ever-increasing cross-border relationships and movements (Castles & Miller, 1998; Portes et al., 1999; Portes, 2001; Kivisto, 2001; Pries, 1997; Vertovec, 1999). Whereas international migration is nothing new and exchanges of economic or cultural goods have been occurring ever since, the pivotal difference of the new era of transnational migration and its impact is the constant and extensive cross-border "back and forth" of exchanges. In contrast to usually limited movements once an international migration (e.g. from the Old to the New World) occurred, the transnational migration is characterized by "sustained linkages and ongoing exchanges among non-state actors based across national borders" (Vertovec, 2009: 3). Such transnational practices are "[f]acilitated, but not caused" (Vertovec, 2009: 2) by the development of internet and communication media as well as innovation of transportation technologies.[1] Migrants and non-migrants alike are connected by their activities "here and there" across national

[1] Whereas the difference between the inter- and transnational migration lies in the focus of the multidirectional and extensive exchanges, debates on the difference to the "global" has also brought some confusion. However, Michael Peter Smith, basing on Michael Kearney's work, (2001: 3) distinguishes clearly between these terms. According to Smith (2001: 3), the global depicts the decentralization from specific national territories and focuses on the insignificance of national borders; the "transnational," however, still "insists on the continuing significance of borders, state policies, and national identities even as these are often transgressed by transnational communication circuits and social practices" (ibid.). The global in this work is therefore understood as a characteristic describing the globe-encompassing feature, whereas transnational as practices and behaviors transgressing conventional national borders.

borders and spanning the globe. The transnationality[2] is distinguished by "its implicit focus on network relations and the idea that transnationality entails the emergence of a new quality of economic, social, and cultural relations both within and between the interacting organizations, communities, territories, and places" (Krätke et al., 2012: 1).

The transnational migrants, also named transmigrants, on a daily basis continue "to participate in the political and economic lives of their homelands, even as they are incorporated into their host societies" and are characterized by "keeping their feet in both worlds" (Levitt, 2001: 3). Apart from slightly different definitions in literature,[3] it can be summed up that transnational migrants are those whose "transnational social relations are 'anchored' in while also transcending one or more nation-states" (Smith, 2001: 3), but also "produce and consume, and so construct their identities" (Kearney, 1995: 233) "in ways quite different from the consumption of 'traditional culture'"[4] (ibid.). Anchoring in specific local cultures and communities, especially both in their country of origin and destination(s), transnational migrants keep attachments to the societies but also often share a specific newly created culture of their own different from each of the local cultures. It should also be noted that transnationalization of societies and communities occur not only within the migrant population. It also involves the local societies as also non-migrants become involved in these daily practices (Faist, 2010).

With the transnational paradigm spreading into a broad field of social and cultural sciences (cf. Khagram & Levitt, 2008), empirical research followed in sociological and anthropological field analyzing the emerging transnational communities (Appadurai, 1996; Levitt, 2001; Ong, 1999) and in more economic contexts transnational ties of entrepreneurs and

[2] Whereas Krätke et al. (2012) use the terms "transnationalism" and "transnationality" interchangeably, there is indeed also a debate on distinguishing these from each other (see Glick-Schiller, 2012). As this is less relevant for the purpose of the monograph, these terms will be used interchangeably here.

[3] See Kearney (1995), Smith (2001), Levitt (2001), Sklair (2001).

[4] This passage refers to the effect and consumption of mass media but can be also understood in a more general term. Kearney, although correct in this respect, further continues to discuss "hyperspaces," in which transnational migrants and communities are being produced and in which they lived in. As geographers we should contest such concept of intangible and diffuse space and attempt to more clearly conceptualize and empirically investigate the matter.

businesses (Saxenian, 2007; Portes et al., 2002; see next section). Initially dominated by North American cases[5] the transnational turn in migration research eventually also reached Europe and other regions on the globe (Faist, 2000). What is important to note in this context is that we here have transmigrants of lower-skilled backgrounds, be it those in guest worker schemes or in other labor migrant contexts. The research on transnational migrants primarily deals with the "transnationalism from below," rightly so, turning the focus on the challenges, struggles but also opportunities and the support networks of migrants. They highlight the everyday practices and the mechanism in which they cope with the migration and migrant experience.

The Transnationalism in Economic Geography

Debates on processes crossing national borders and questioning the relevance of nation-states as a basis for international relations and global issues, which appeared new in migration research, had been actually discussed and already used as a concept much earlier in international business studies and economic geography. Whereas "transnational" in social sciences, in particular migration studies, stands for the general shift in migration trends of the current era, and also signifies a groundbreaking paradigm shift in the studies of migration in general, the terminology within economic and business contexts is more concrete.

The term, as it is still used, refers to the specific corporation types and business operations. While all multi-, inter- and transnational corporations are characterized by organizational processes and economic activities which cross national borders, Bartlett and Ghoshal (2002: 57ff.) give clear economic definitions for the different organizational models. *Inter*national signifies politics and processes that are carried out or experienced *between* nation-states. *Multi*-national are processes where *multiple* nation-states are involved in (rather irrelevant for the debate on transnational migration though having significance in migration policies). Focusing on such corporate organizational structures, the

[5] Kearney (1995), Roberts et al. (1999), Levitt (2001), Guarnizo (1997), cf. Smith and Guarnizo (1998).

transnationality in this context, in contrast to international and multinational corporations, lies in the locational dispersion and in the multidirectionally interconnected coordination and control of business processes within a corporation (see Table 2.1).

Dicken defines the transnational corporation as “a firm that has the power to coordinate and control operations in more than one country, even if it does not own them” (Dicken, 2007: 106). Spatially, as he continues, they are active “both within and between different countries,” yet “within production networks” (ibid.) and showing “potential geographical flexibility” (ibid. 107). Such economic definition seems to imply that corporations operate irrespective of national borders and that through their constitution of shareholders, executives and business partners from all over the globe, they are increasingly detached from nation-states. Whereas there might be a tinge of truth in the societal impact of such global and transnational operations, in the mindset of the professionals involved (as such universalist perspectives have been pushed forward by human resource managements in transnational corporations; cf. Brewster, 2007), the actual business reality is that trades, taxes and other

Table 2.1 Cross-border corporate operations according to Bartlett and Ghoshal (2002: 57)

Ogranizational characteristics	Multinational	International	Transnational
Configuration of assets and capabilities	Decentralized and nationally self-sufficient	Sources of core competencies centralized, others decentralized	Dispersed, interdependent, and specialized
Roles of overseas operations	Sensing and exploiting local opportunities	Adapting and leveraging parent company competencies	Differentiated contributions by national units to integrated worldwide operations
Development and diffusion of knoledge	Knowledge developed and retained within each unit	Knowledge development at the center and transferred to overseas unit	Knowledge developed jointly and shared worldwide

transactions necessary for operations are—despite agreements on supranational level—still heavily anchored in national policies.

The debate on such question whether nation-states and their borders still have any significance in today's world has also arisen in the context of transnationalism research as many political, economic and social actors appear to carry out their activities irrespective of such constraints.[6] Especially aforementioned multi-national or transnational corporations appear to be independent in their activities from the borders.[7] However, it can be maintained that the existence of nation-states themselves and also their role as the regulatory framework for any interactions, in which actors including transnational migrants are embedded, have been widely recognized and accepted. Transnationalism debates are primarily about the nature of migration, including migrants' everyday lives, which are embedded in the constant cross-bordering beyond national states, and less about the idea of conceptually depriving the national states their function and existence.

2.2 The Vacuum of Transnational Professionals

With these different strands on transnationalism, there is a group of people who have been somewhat overlooked: the transnational migrants working in and for transnational corporations. Named transnational professionals, corporate migrants or often just as "expats" (short for expatriates), the fact that they are part of an ever-increasing group of transnational migrants appears to be ignored. Transnational professionals are objects of

---

[6] The theoretical debates, which were in fact the initial grounds for the transnationalism debates, contest the concept of nation-states and also touch upon issues of nationality, nationalism and citizenship in both political and social contexts (see more in Anderson, 2006; Basch et al., 1994; Glick-Schiller et al., 1992; Wimmer & Glick Schiller, 2002), yet discussion of these debates is not directly relevant for the purpose of the narrative of the book and will not be further explicated.

[7] In some literature, the term "trans-border" has been also used with regard to transnational migration. However, transborder studies originate in the tradition of political and historical sciences and also focus more on the border itself and the actual crossing of borders. Although there is interchangeable usage in literature, as "transborder" focuses less on the societal effects of the transnational nature of migration, this terminology will not be used here.

research but primarily on the side of economic geography and international business studies. The focus of these studies is the involvement of these professionals in transnational business networks and their functions in transnational corporations. Literature on the assignment of such personnel as part of corporate strategy and human resource management is in abundance since the prevalence of operational transnationalization of companies in the 1980s.

Although identified as being actors in the global economy in such perspectives, their actual role and existence as migrants themselves are neglected. In fact, it is only in recent years through the reflexive turn in migration research (Amelina, 2021) that the problematization of migrants, the discourse on migration as it is including categories of migrant differences have started being questioned (Dahinden, 2016; Schapendonk, 2021). With only a few exceptions, such as an overarching discussion of these transnationals as part of a globally engaged transnational capitalist class (Sklair, 2001) or superrich elites (Beaverstock, 2004, 2012, 2018), who differ from transnational entrepreneurs (Saxenian, 2002, 2007; Yamamura & Lassalle, 2022), have been a group of transmigrants[8] overlooked. Although research interest is growing, empirical research on transnational corporate professionals' individual social practices[9] and everyday lives as part of the transmigrant debate is still rare. Research on transnational professionals is existent in social sciences but focus on the economic geographical perspective on business communities and practices (Carroll & Fennema, 2002; Morgan, 2001) or their role in transnational corporate networks (Beaverstock, 2004; Faulconbridge, 2007, 2008). In these research studies, transnational professionals and their interactions in the form of business communities and networks are contextualized in the management and operations of

[8] Not necessarily part of the transnational migration research but affiliated are research works on the consequences to the children's generation as part of a discussion on so-called third culture kids—who are children growing up in mobility through their parents' frequent and extensive overseas assignments (Pollock & Van Reken, 2001)—debates (Goulbourne et al., 2010; Waters, 2003).

[9] Although acknowledging the growing interest in practice theories (Reckwitz, 2002; Schatzki, 1996; Schatzki et al., 2001) in human geography, particularly within the new cultural geography, "social practices" or "everyday practices" in this monograph is used as in the social scientific and anthropological tradition of migration research and does not refer to the "practice turn" as further theorization of practices would not be possible within the scope of the presented research.

transnational firms. The research does not reach the subjective level of transnational professionals as migrants themselves outside of their functions in their respective working environment. It is only recently in the debates on so-called privileged migration that these transnational professionals have come into focus as migrants themselves.[10]

As a matter of fact, the perception and identification of transnational corporate professionals as migrants is paradoxical in society. Generally both transnational professionals themselves and also people dealing with these "foreign residents," such as real estate agents or metropolitan governmental officials, are surprised to be asked about them as migrants. There is a notion of temporality and privilege bound to migrants of transnational corporations and other highly skilled occupations, whereas indeed "migrants" per se tend to be associated strongly and intuitively with lower-skilled and maybe even "colored" migration. As an article in the *Wall Street Journal* depicts below, this bias naming and perceiving primarily white and highly skilled migrants as "expats" in the global city Hong Kong is ubiquitous:

> So it's strange to hear some people in Hong Kong described as expats, but not others. Anyone with roots in a Western country is considered an expat. But the distinction is muddied among Hong Kong's deeply entrenched Southeast Asian community. Filipino domestic helpers are just guests, even if they've been here for decades. Mandarin-speaking mainland Chinese are rarely regarded as expats, but they are certainly not locals. By contrast, a native Cantonese speaker earns an automatic right to belong, even if she spent most of her life in Sydney or Vancouver. (…)
>
> It depends on social class, country of origin and economic status. (…) A more current interpretation of the term 'expat' has more to do with privilege. Expats are free to roam between countries and cultures, privileges not afforded to those considered immigrants or migrant workers. (DeWolf, 2014)

This societal bias of perception is definitely an interesting argument to follow in the context of postcolonialism and even critical whiteness research. It coincides with the disciplinary divide on transnationalism

[10] Cf. Croucher (2012), Cranston and Lloyd (2019), Duplan and Cranston (2023), Kunz (2016).

research as outlined previously. Without the transnational professionals and the general public perceiving them and themselves as part of the migrant population despite their legal and societal status as being so, and with the divergent foci on the phenomenon of diversification in society in the disciplines, a vacuum around transnational professionals as transnational migrants has been created.

As the journalist of the article indicates, "the nomenclature is outdated, rooted in a time when voyages involved a one-way ticket on a steamship" (ibid). This may at first sight appear trivial and colloquial, but the idea can be transferred to the recent debates in migration research, namely, to the concept of superdiversity. The superdiversity as coined by Steven Vertovec is a new approach to do justice to the complexity of the global society dealing with "a dynamic interplay of variables among an increased number of new, small and scattered, multiple-origin, transnationally connected, socio-economically differentiated and legally stratified immigrants" (Vertovec, 2007: 1027). The migrant population has been diversifying with regard to factors such as the ethnic, religious, national, cultural and religious origins, the migration channel through which the groups enter the migration paths, which again can be multiple along the lifecycle, the legal status and the individual's human capital, by that also access to employment. Then, the flows of migrants also differ in the locational perspective as well as in their social practices of transnationalism beyond the national borders, but also the destination societies have "usually chequered responses by local authorities, services providers and local residents (which often tend to function by way of assumptions based on previous experiences with migrants and ethnic minorities)" (ibid: 1049). All these dimensions contribute to a new complexity in the global society.

In fact, what the transnational migration debates additionally miss is the growing complexity also in transnational migration. The common notion in current research is basically twofold. One is that transnational migration patterns are seen in a bidirectional context, meaning a frequent and daily back and forth of social practices between the country of origin and country of destination (Portes et al., 1999; Smith & Guarnizo, 1998). Such limited understanding of transnational migration is also reflected in the notion of "sending" and "receiving" countries or in the dichotomous idea of "country of origin" and "country of destination." The other critique that needs to be voiced is that although there are multi-sited perspectives on transnational migration, such cases are still viewed through a mono-ethnic lens. Often even using term of diasporas,[11] such transnationalism studies (e.g. Voigt-Graf, 2004) have the multilocational migrations of one single ethnic group in focus, adopting an ethno-focal lens.

These bidirectional or mono-ethnic forms of transnational migrations are undeniably the most prevalent type of migration until very recently. However, the migrant population of those moving to multiple destinations and not only between one place of origin to another place of destination over their life cycles is growing every moment. As a thought exercise, if one imagines an international student who leaves his or her country of origin, and then looks for job opportunities in countries beyond the study country, this migrant already becomes outside of the box for aforementioned transnational migration research. The same applies to the generation of high school graduates who go abroad for a year for work and travel, do an international year or even the graduate studies elsewhere and depart on a career in an internationally operating company, or those retirees who decide to move to a sunnier place. Transnational migration has become more common than we might imagine at first.

[11] Diaspora or diasporic communities have been partly used interchangeably in the debates on transnational migrants, and even one of the most prominent diaspora researchers mentions this term and global cities in the same breath (Cohen, 1997) and also intermingles further terms, such as denizens, cosmopolitanism and transnational, in an attempt to virtually stretch the original idea of diaspora to a universal synonym for global migration. Cohen even proposes "the good gardener's guide to diasporas" encompassing not only victim/refugee, imperial/colonial types but also creating labor/service, trade/business/professional and cultural/hybrid/postmodern types of diasporas. Such notion of undifferentiated and ubiquitous usage of the terminology, which Brubaker has mocked as a "the 'diaspora' diaspora" (Brubaker, 2005), is not adopted in this work. Diaspora may not need to be narrowly restricted to the original or classical case of Jewish migration, but at least should be reserved for those phenomena of migration where the migrants are being dispersed rather than deliberately disperse themselves around the globe. There are surely different criteria discussed in diaspora literature (Brubaker, 2005; Braziel & Mannur, 2003; Dufoix, 2008), yet what has made the diaspora historically significant appears to be the aspect of dispossession and involuntariness, and should not be confused with the transnational migration, especially of those privileged professionals who are voluntary labor, if not even part of lifestyle migrants (Benson and O'Reilly 2009, 2012).

In addition to the multidirectionality, the superdiversity shift in migration also calls for a more differentiated view on the complexity of the diversity of the migrant population (Vertovec, 2007). The times in which it could be criticized that the overall numbers of transmigrants are too little to attach importance to (Haug, 2000; Pries, 2005) has long passed. Also, as Levitt et al. (2003) point out, even though the transmigrants may themselves be small in number, transnational practices are shared by many people and also involve non-mobile people (Faist, 2010; Guarnizo et al., 2003; Vertovec, 2009). As much as the idea of transnationalism from below that "labor stays local, whereas capital ranges global" (Portes et al., 1999: 227) is pointed out correctly, such a unidirectional approach misses out that the transnationalization drive is not directed by ghost hands of the global capital or global economy but by individuals with their own interests and practices, too. The agency of individuals in issues of mobilities also echoes in the growing research on lifestyle migration (Benson & O'Reilly, 2012), which subsumes different types of (privileged) migration for the purpose of self-realization and/or the pursuit of a better quality of life.

The approach of superdiversity within migration research tackles these conceptual shortcomings. Also, what the superdiversity debate also emphasizes is the need to take an urban lens to the studying of the social phenomenon of transnational migrants (Meissner & Vertovec, 2015). Cities, in particular global cities, are places where a tremendous variety of migrants arrive, reside and move, and, thus, such complexity of superdiversity especially becomes visible in the urban landscape (Yamamura & Lassalle, 2020; Yamamura, 2022a, 2022b). Consequently, the following section gives an overview on the nexus of transnational migration and cities, with the link to more recent research on superdiversity.

2.3 Transnational Migration and Cities

Transnational Urbanism

The concept of "transnational urbanism" was coined by Michael Peter Smith (2001, 2005) and such urban lens still remains the main approach to studying transnationalism. Cities are put into focus and the urban lens is taken as the approach toward better understanding transnational practices by migrants. Cities, according to Smith, are:

> sites of transnational practices, contexts of transnational network formation, socially structured settings for social interaction, and mediators of the power, meaning, and effects of transnational flows 'from above' as well as 'from below.' (Smith, 2001: 19)

Further debates on transnationalism and cities have also brought forth the concept of "translocality" (Smith, 2001) which emphasizes even stronger the multilocality of social ties and social practices relating to multiple places on a transnational context. Translocality, as Wehrhahn and Tölle (2016) write, emphasizes the simultaneous presence and co-presence of migrants at multiple sites. Bringing different studies on translocal geography together, Brickell and Datta point out the "situatedness across scales," respectively the "multiscalar situatedness of migrants' experiences" (2011: 11) in this context. In the strict sense of the understanding of translocal geography, transnational practices are not only bound to urban spaces. The concept of keeping simultaneously intensive social ties and practices by migrants is not limited to urban space (cf. also Greiner & Sakdapolrak, 2013). As Basch et al. (1994) point out, national borders are not crucial for the constitution of transnational social spaces. Other types of borders, such as other geographical, ethnic or cultural, can also have the same implications. For this reason, terms such as plurilocality have also emerged (Ossenbrügge, 2005; Pries, 1999).

Although there are, generally speaking, plausible grounds for the usage of alternative terminology (with slight connotational or conceptual differences and emphasis), such as plurilocality or translocality, the focus of this monograph, that is, the transnational professionals in global cities, are clearly characterized by their embeddedness in transnational corporate contexts. Through such affiliation and contextualization in transnational corporations, the spatial dimension is then primarily focused on the urban level, that is, global cities because these are the sites where transnational corporations span their global networks. Against this background, the transnational professional migrants and therefor our debate here can and should be discussed in the context of transnational urbanism rather than translocality or plurilocality.

The multi-scalarity mentioned above can also be observed not only on supra-urban scales, but also within the city. Brickell and Datta (2011: 13ff.) bring together studies on translocal homes, translocal neighborhoods and translocal cities. The conceptualization of the geography of scales in the context of transnationalism and cities is quite complex. Krätke et al. propose a "basic classification of the manifestations of transnational urban spaces" (2012: 10). They discuss that such urban spaces can be distinguished by different levels of analysis, whereas they differentiate between the economic transnationalism and socio-cultural transnationalism at each level. If looked at from the macro-level, cities are seen as part of the global urban system, on the meso-level the city and its urban region and, on the micro-level, the transnational urban space is analyzed at the level of particular urban districts and areas. Furthermore, to accommodate the translocal aspect of the debates as outlined above, the authors also list up the "specific localities within the city" (ibid.) as a further unnamed level below the micro-level.[12]

As the transnational urban spaces of transnational professionals are in fact multi-scalar, the overall approach of the macro-, meso- and micro-levels will be adopted in this present work. The multi-scalar approach is the actual geographical perspective on the diversification of urban spaces that have been discussed in adjacent research, such as entrepreneurial superdiversity or intersectionality or in larger migration and integration

[12] In each case of level or geographical scale, the differentiation between economic and socio-cultural transnationalism results in a categorization of different actors, whereas these repeat in the table. The classification shows some inconsistency in that it brings in non-spatial sociological concepts of "social fields." The strict dichotomous differentiation between transmigrants and global elites also appears to neglect the individuality or dual role of such global elite, respectively the transnational professional migrants 'from above' appear to be neglected.

policy contexts (Yamamura, 2023; Yamamura et al., 2022). Translating into the global cities context, the spatial diversification will be analyzed from the local urban level of Tokyo, then from the meso-level the transnational space's extent to the global city-region and finally its dispersion over the global cities network.

Global Cities

Following the debates on transnational urbanism, global cities have emerged as a special arena for the study of such transnational activities and practices. As Friedmann set out in his original hypothesis on global cities as early as in the 1980s (1986: 58), the "form and extent of a [global] city's integration with the world economy" including its control and command functions and its role in the division of labor as Sassen later conceptualizes (1991), "will be decisive for the structural changes occurring within the city" (Friedmann, 1986: 71). As the global city theory itself is not the focus of this monograph, but its urban context for the study of spatial diversification by transnational migrants, an extensive literature review on its economic functions will be omitted here; the focus lies on the nexus of transnational migrants and cities in the global city context.

Global cities have been researched extensively ever since the anteceding term and concepts of the world cities have been widely disseminated already in the late 1960s (Hall, 1966) and more prominently in the context of world city hypothesis in the 1980s (Friedmann, 1986; Friedmann & Wolff, 1982). Saskia Sassen then coined the term of global city through her seminal work in 1991 (Sassen, 1991; see Brenner & Keil, 2006; Derudder, 2012 for more details on development, theories and concepts on global cities). The main idea of the global city theory is the dominance of specific cities in the global economy which function as control and command centers, where transnational corporations, in particular of the FIRE sector are concentrated. Originally discussing global cities as the location of headquarters of multi-national and transnational corporations, the concept evolved into focusing on service providers for the FIRE sector. The debate was followed by further theoretical but also and more importantly empirical work on the interconnectivity and networks of these global cities (e.g. Knox & Taylor, 1995; Taylor, 2004). What has come more into recent focus in the research and understanding of the global cities and global city-region are the interconnectedness of the cities and the establishing of a global urban system (Derudder, 2012; Taylor et al., 2011) as well as its connectedness to global production networks, respectively global commodity chains (Derudder & Witlox, 2011). With the exceptions on research on the actual business- and global cities making through corporate activities, which base on qualitative research (Parnreiter, 2012), the majority of recent global city research appears to be based on quantitative data[13] and to be focusing on policy issues (cp. Taylor et al., 2011), fading the actual societal aspects more into the background.

Initially, debates on issues of social polarization emerged, questioning the global city theory as laid out by Sassen in its contents (Hamnett, 1994; Fainstein, 2001; cf. debate between White, Smith and Sassen in Urban Affairs Review in 1998). With regard to the nexus of the migrants and city, Sassen sets out that global cities is where highly skilled migrants from the corporate elite but also low-skilled migrants to serve the class come together. Global cities have been identified as being "the most strategic spaces for the formation of transnational identities and communities" (Sassen, 1996: 205) and where "transnationality from above emerges as a distinct physical and social environment" (Parnreiter, 2012: 100). However, apart from quantitative data on population—which not in all cases of global cities, particularly not in the case of Tokyo, did not even fully reflect this trend (Fainstein, 2001, 2011)—empirical research on this supposed phenomenon of dual migration into global cities, especially in its impact on spatial diversification, did not follow. As mentioned before, the focus on the actors in global cities concentrated on the transnational corporations and their professionals, and rather neglected these professionals in their character of being transmigrants. Despite the empirical negligence of global cities in the transnational migration research, the following quote illustrates its potential as a research field:

---

[13] As noted before, the different focus has its roots in the relational understanding and contextualization of the city in case of the more economic geographical approach in global cities research.

> Global cities are, then, thick enabling environments for these [transnational] types of activities, even though the networks themselves are not urban per se. In this regard, these cities help people experience themselves as part of global non-state networks as they live their daily lives. They enact global civil society in the micro-spaces of daily life rather than on some putative global stage. (Sassen, 2011: 87)

2.4 Excursus on Global City Tokyo

Tokyo, as a particular global city has been neglected—or somehow recently dropped—from recent debates on global cities. As presented above, the primary aim of much of studies on global cities conducted by economic geographers and urban researchers is to find and explain the relational interconnectedness between those cities, and therefore, mostly rely on statistics and quantitative data (Derudder & Taylor, 2016). With some notable exceptions (cf. Parnreiter, 2012), qualitative empirical data are rather scarce in research on glocal cities, leading to a lack of in-depth exploration and analysis of socio-spatial patterns of actors in the global cities. Other research in urban sociology or from anthropological perspectives have been using the term "global" in a more generic way while researching within the global city contexts, but have not yet used the concept or the theoretical framework of global cities. Going back to global city Tokyo, despite being classified among the "big three" along with New York and London and being considered as one of the most established global cities (Sassen, 1991), has remained empirically under-researched, even more specifically regarding international migration. The rise of other vibrant and economically dynamic global cities in the Asian region (with fastest economic growth), as, for example, Hong Kong and Singapore, explains the shift of general academic interest toward those new global cities. Also, the long-lasting economic downturn in Japan since the early 1990s, known as the "lost decades" has additionally discouraged researchers' interest in the global city Tokyo. The lack of empirical work on Tokyo as a Global City might also been explained by methodological issues related to field access as it might indeed appear more difficult to engage in qualitative fieldworks in Tokyo more than in other bilingual cities—such as those previously mentioned—due to Japanese language.

More generally, there has been increasing skepticism regarding the global city status of Tokyo especially with regard to "lost decade(s)" of economic downturn. Already in the 1980s, John Friedmann classified Tokyo as a "major atypical case" regarding "the concentration and accumulation of international capital" (Friedmann, 1986: 69), leading to further reservations about Tokyo's position as a Global City. Further contributions following the seminal works of Friedmann (1986) and Sassen (1991), while still in the midst of the bubble economic era, or still relying on economic data from this era, confirmed the position of Tokyo as a world city (Douglass, 1988, 1993; Fujita, 1991; Machimura, 1992, 1994; Rimmer, 1986). However, the specific—or the "atypical"—nature of Tokyo was many times compared to other major global cities (Kamo, 2005; Knox & Taylor, 1995; Machimura, 1994). Adding to Friedmann's (1986) original comment on the limited accumulation of international capital in Tokyo, there was growing skepticism about Tokyo's qualification and classification as a world city.

First, Tokyo's peculiar urban politics and urban development policies took place in the context of academic debates on Asian anomaly of the governmental state (White, 1998; Hill & Fujita, 2000; Douglass, 2000; Kamo, 2000; Hill & Kim, 2000; Saito, 2003; Saito & Thornley, 2003). Second, the state of the financial industry in Tokyo was considered "immature"—at least in comparison with other Alpha + or Alpha++ global cities which can be classified as major financial centers such as New York or London—whereas Tokyo's competitiveness was maintained by manufacturing industry (Fujita, 1991; Johnson, 1982; Kamo, 1988). Third, unlike other global cities, Tokyo is characterized by limited immigration, which led to questioning on the status of Tokyo as a "top tier" global city (as portrayed by the hefty exchange in Urban Studies 1998 between White, Sassen and Smith). Finally, the symbolic consequence of the long-lasting economic downturn and the accompanying post-bubble economic malaise of the "lost decades" in politics was the abandon of the euphoric usage of the notion of "world-leading global city" Tokyo (cf. Machimura, 1998).

In spite of such skepticisms and growing doubts about Tokyo's status as a global city, data from the GaWC research network (Taylor et al., 2011; the World According to GaWC, 2020) show evidence that Tokyo—even though now an Alpha + behind the two Alpha ++ cities of New York and London—has not actually really dropped much in terms of interconnectivity in absolute terms but faces increasing competition from growing Asian global cities. However, Tokyo retains a position among the most interconnected global cities (Derudder & Taylor, 2016) and, thus still remains an interesting global city to research on.

In that sense, global city Tokyo is a relevant location in which to explore the nexus of transnational migrations and cities. Considered as part of the wider context of (super)diversification occurring in global cities, the dynamics induced by the flows of transnational migrants in global cities needs to be reconsidered. Global cities do not only remain in the network of peer cities after developing such economic ties. Those inter-relational ties also show dynamics within the networks. Since the early 1990s, the global cities of New York, London and Tokyo have been considered as a top tier; however, the global network of global cities has expanded and the emergence of cities from the Asian region, such as Hong Kong and Singapore, as mentioned previously, increasingly became destinations for foreign direct investments, and, consequently for research in academia. The rise of those cities in the networks of global cities has had consequences on academic debates with regard to the nature of their development as global cities. By comparison to the Anglo-American and European global cities, these emerging Asian global cities present different characteristics regarding their diversity. As postcolonial and bilingual cities, they also differ from other non-English-speaking global cities that are emerging in other regions.

The choice of Tokyo as a location for this research on spatial diversification in global cities thus finds justification in the following rationales. First, Tokyo is considered as a top tier global city since the origins of the concept (cf. Sassen, 1991) and has remained as such in spite of skepticism on its "atypical nature" (Friedmann, 1986), still being ranked as an Alpha+, just behind New York and London in the latest ranking (Derudder & Taylor, 2016). Despite this position among the major global cities, Tokyo has attracted little interest and remains under-researched empirically, probably partly because of linguistic barriers, in contrast to the abundance of empirical work conducted in New York and London. Second, the emergence of global cities in Asia, starting with Hong Kong and Singapore but also Shanghai and Beijing, contrast with the long downturn faced by the Japanese economy during the "lost decades," and comparatively attract further interest of economic actors and academic researchers. However, this disinterest leaves open questions on the socio-spatial diversification phenomenon occurring within the current global city of Tokyo. Third, Tokyo is still a unique case among global cities. Being at the same time a non-English-speaking city but also a non-postcolonial location, it presents distinctive characteristics and thus an interesting case regarding the future of global cities. By not following the steps of major global cities such as New York or London in terms of migration and urban diversification, Tokyo is an interesting case—if not a prototype—that might well inform researchers on the development of new and emerging global cities, such as in Arab or Southeast Asian region.

2.5 Socio-spatial Diversification in Global Cities

The concise summary of the different debates on transnationalism as well as on transnational urbanism have clarified that transnational professionals especially in the context of global cities are the key to understanding the spatial diversification in times of increased transnational migration. Global cities are the spaces where the dichotomy of transnationalism from above and below comes together. Moreover, the constitution of global cities themselves as embedded in the structures of the global economy, but also as those creating the network of centers in the global economy, brings a complex construct to the transnational space.

The complex interplay of the transnational space is pointed out by Faist as follows:

> Transnational social spaces are constituted by the various forms of resources or capital of spatially mobile and immobile persons, on the one hand, and the regulations imposed by nation-states and various other opportunities and constraints, on the other; for example, state-controlled immigration and refugee policies, and institutions in ethnic communities. (Faist, 2000: 200)[14]

In-between these different structures in which transnational migrants are embedded and also contribute to the structures themselves, global cities are where, as Pries points out,

> pluri-locally spanned transnational social spaces as social realities and entities that grow up either from the grassroots by international migration or through a complex top-down and bottom-up process brought about by international business companies emerge. (2001: 3)

At the same time, regarding the actual daily lives of people in such cities, Ossenbrügge also uses the term of "glocal spaces" to characterize such spaces (2005).

> Individuals with very different lifestyles and social relations can live in immediate proximity. Persons with distinctly local rootedness and cosmopolitans with intensive cultural relations to other (world) regions form a neighborhood in global places. (2005: 27; author's translation)

bringing together very local and global aspects of social activities including activity spaces into the same locations.

These individuals such as transnational professionals then, in social and economic interactions with also local immobile residents (Faist, 2000), contribute to an increasing social but also spatial diversification in global cities. The transnational social space consists not only of the migrants themselves, but also involves naturally the local population, too. With the diversity then of not only the transnational professionals

[14] This approach should be contextualized into the debates on mixed-embeddedness (Kloosterman et al., 1999), in which the multiple embeddedness of migrants both in social context within structures but also in institutional is being discussed.

themselves, who are not of single ethnic origin, but of multicultural and multiethnic backgrounds as a social group, further diversity is added by the "other" transnational migrants from below who are also attracted to the city, especially in the case of global cities.

Such complexity of overlapping and interplays of different factors in society, in particular of the variety of factors surrounding the migrant population, including the country or origin, linguistic, religious, ethnic backgrounds, but also legal status and access to employment, etc., was designated as "superdiversity" (Vertovec, 2007). The transnational space, which is "complex, multidimensional and multiply inhabited" (Crang et al., 2004: 3), is then the optimal research field for the study of spatial diversification.

While superdiversity already at the time of its inception by Vertovec in 2007 referred to the migrant-led diversification in cities such as London, the meaning of the urban lens has been repeatedly emphasized within superdiversity research (Meissner & Vertovec, 2015; Yamamura et al., 2022). As most of the research on urban diversity emerged from anthropological research, the focus is less on the spatiality of diversity in cities itself but more on issues of conviviality and social encounter in mixed cities or neighborhoods, while not necessarily using the concept of superdiversity but depicting the encounters of differences in urban space.[15] When it comes to spatial issue, concepts, such as "commonplace diversity" (Wessendorf, 2013, 2014) or "differential inclusion" (Ye, 2017; Yeoh, 2006) have emerged. They describe and theorize, on the one hand, how "ethnic, religious and linguistic diversity [is] experienced as a normal part of social life and not as something particularly special" in convivial urban spaces (Wessendorf, 2013: 407). On the other hand, they discuss how different migrant groups are included differently within the city, such as Singapore, referring to the social and administrative differences in migrants (Ye, 2017; Yeoh, 2006). As it was previously pointed out (Yamamura, 2022a, 2022b), the importance of social spaces is identified in such work, that is, in Wessendorf, 2013: 410:

[15] Cf. Heil (2020), Valentine (2008), Vincent et al. (2017), Wessendorf (2013, 2014), Wilson (2017), Ye (2017).

> social spaces [...] play an important role in the process of familiarisation with people who are different and in getting accustomed to communicating across difference.

Yet, the urban superdiversity remains rather abstract and the actual "spatiality of superdiversity" (Yamamura, 2022b) under-researched.

Without diverging into details of different urban research strands discussing related phenomena to such issues of diversification, a noteworthy line of research is that on planetary or transnational gentrification.[16] As the term of gentrification already alludes, transnational migrants play a key role in contributing to the transformation of urban spaces locally and in the global rent gaps debates. However, the focus is primarily on "re-spatialisation of global inequality" (Hayes & Zaban, 2020: 3010) at the planetary level and the question on the actual differential superdiversification in the urban space remains unanswered. Also, historically studies on gated communities (Atkinson & Blandy, 2013; Blakely & Snyder, 1997) or on the concept of upper-class citadels (Marcuse, 2000) have discussed upper-class residents and often also of upper-class migrants in city spaces, but the socio-spatial differentiations within such cities beyond the physical characteristics of the urban space are still rare, let alone those discussing the spaces of transnational professionals in global cities.

In this context, Pries should be mentioned as he discusses the "stackedness" of transnational spaces in global cities, which will become relevant later when the different transnational spaces in Tokyo will be discussed:

> A global city is an agglomeration of distinct social spaces existing within the same geographic area. These 'stacked' social spaces do not necessarily interact intensively with each other (for example ethnically and culturally distinct groups of migrant labourers). These social spaces could have stronger ties with social spaces in geographic spaces other than their own. Therefore, in addition to different social spaces being 'stacked' within the same geographic space, a social space can also expand over several geographic spaces. (Pries, 2001: 5)

[16] Hayes and Zaban (2020), Lees et al. (2016), Sigler and Wachsmuth (2016).

Recapturing the different strands on transnationalism, the city and the migrants, Fig. 2.2 summarizes the context of this monograph. The focus on the transnational (urban) space brings together the different strands on the global cities networks with their transnational corporations (TNCs) and transnational (TNPs) and the transnational migrant networks on the side of transnational migration research, when transnational professionals are viewed as part of transnational migrant population. In consequence then, the overall spatial diversification caused by the increased transnational migration in the global society can then be identified.

This chapter has now set the scene for the overall narrative of the monograph, locating the research topic within the ongoing debates surrounding the increasing transnationalism in the global society and their impact on the urban spaces. It has become clear that the research into, first, transnational professionals as a specific group of transnational migrants, and, second, the transnational space in particular in the context of the global cities network, bears strong potentials in better understanding the dimension and extent of the spatial diversification occurring

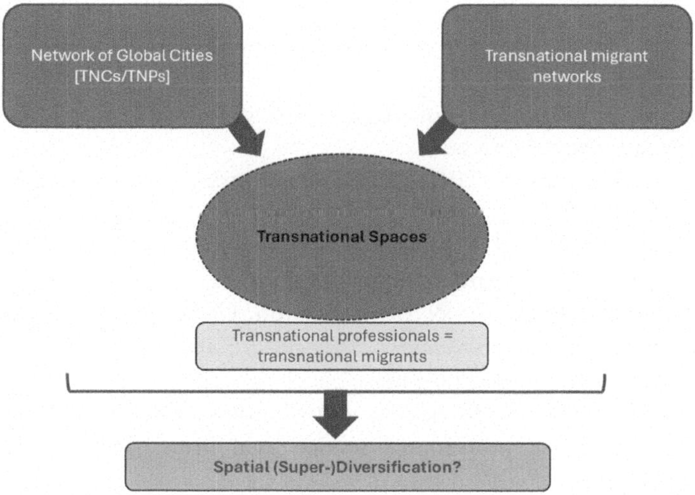

Fig. 2.2 Transnational space of transnational professionals and spatial diversification (author's own)

throughout the world. In fact, the socio-spatial dynamics of transnational professional's urban spaces is the key also to holistically understand and conceptualize the socio-spatial diversification within the discussion of the emerging debates around urban superdiversity.

In this context, the much neglected global city Tokyo provides an interesting field as a city well established as a global city yet empirically under-researched. Tokyo does not only give an interesting insight as it has not been thoroughly empirically investigated within the debate on global cities. It is also a city that is nowadays experiencing an increase in its migrant population and attempting to navigate back to the global forum of leading cities. By looking into a city which appears to be dwindling away in the debate on competitive global cities can be the key to understanding the socio-spatial dynamics of the socio-spatial diversifications in a larger spatiotemporal context.

References

Amelina, A. (2021). After the reflexive turn in migration studies: Towards the doing migration approach, Population. *Space and Place, 27*(1), 1–35. https://doi.org/10.1002/psp.2368

Anderson, B. (2006). *Imagined communities: Reflections on the origin and spread of nationalism* (Rev ed.). Verso.

Appadurai, A. (1996). *Modernity at large. Cultural dimensions of globalization.* Minnesota University Press.

Atkinson, R., & Blandy, S. (2013). *Gated communities: International perspectives.* Routledge.

Bartlett, C. A., & Ghoshal, S. (2002). *Managing across borders: The transnational solution* (2nd ed.). Harvard Business School Press.

Basch, L., Glick Schiller, N., & Blanc-Szanton, C. (1994). *Nations unbound: Transnational projects, postcolonial predicaments, and deterritorialized nation-states.* Gordon & Breach Publishers.

Beaverstock, J. V. (2004). 'Managing across borders': Knowledge management and expatriation in professional service legal firms. *Journal of Economic Geography, 4*(2), 157–179.

Beaverstock, J. V. (2012). The privileged world city: Private banking, wealth management and the bespoke servicing of the global super-rich. In

International handbook of globalization and world cities (pp. 378–389). Edward Elgar.

Beaverstock, J. V. (2018). New insights in reproducing transnational corporate elites: The labour market intermediation of executive search in the pursuit of global talent in Singapore. *Global Networks, 18*(3), 500–522.

Benson, M., & O'Reilly, K. (2012). *Lifestyle migration: Expectations, aspirations and experiences*. Ashgate.

Benson, M., & O'Reilly, K. (2009). Migration and the search for a better way of life: A critical exploration of lifestyle migration. *The Sociological Review, 57*(4), 608–625.

Blakely, E. J., & Snyder, M. G. (1997). *Fortress America: Gated communities in the United States*. Brookings Institution Press.

Braziel, J. E., & Mannur, A. (2003). *Theorizing diaspora: A reader*. Blackwell.

Brenner, N., & Keil, R. (2006). *The global cities reader*. Routledge.

Brettell, C., & Hollifield, J. F. (Eds.). (2000). *Migration theory: Talking across disciplines*. Routledge.

Brewster, C. (2007). Comparative HRM: European views and perspectives. *The International Journal of Human Resource Management, 18*(5), 769–787.

Brickell, K., & Datta, A. (2011). *Translocal geographies*. Ashgate Publishing Ltd.

Brubaker, R. (2005). The 'diaspora'diaspora. *Ethnic and Racial Studies, 28*(1), 1–19.

Carroll, W. K., & Fennema, M. (2002). Is there a transnational business community? *International Sociology, 17*(3), 393–419.

Castles, S., & Miller, M. (1998). *The Age of Migration*. International Population Movements in the Modern World. Red Globe Press London.

Child Hill, R., & Fujita, K. (2000). State restructuring and local power in Japan. *Urban Studies, 37*(4), 673–690.

Cohen, R. (1996). *Theories of migration*. Edward Elgar.

Cohen, R. (1997). *Global diasporas: An introduction*. University of Washington Press.

Crang, P., Dwyer, C., & Jackson, P. (2004). *Transnational spaces*. Routledge.

Cranston, S., & Lloyd, J. (2019). Bursting the bubble: Spatialising safety for privileged migrant women in Singapore. *Antipode, 51*(2), 478–496. https://doi.org/10.1111/anti.12433

Croucher, S. (2012). Privileged mobility in an age of globality. *Societies, 2*(1), 1–13.

Dahinden, J. (2016). A plea for the "de-migranticization" of research on migration and integration. *Ethnic and Racial Studies, 39*(13), 2207–2225. https://doi.org/10.1080/01419870.2015.1124129

Derudder, B. (Ed.). (2012). *International handbook of globalization and world cities*. Edward Elgar Publishing.

Derudder, B., & Taylor, P. (2016). Change in the World City Network, 2000–2012. *The Professional Geographer, 68*(4), 624–637. https://doi.org/10.1080/00330124.2016.1157500

Derudder, B., & Witlox, F. (2011). *Commodity chains and world cities*. John Wiley & Sons.

DeWolf, C. (2014). *In Hong Kong, just who is an expat, anyway?* Dec. 29, 2014 11:56 pm ET. Retrieved September 15, 2023, from https://blogs.wsj.com/expat/2014/12/29/in-hong-kong-just-who-is-an-expat-anyway/

Dicken, P. (2007). *Global shift: Mapping the changing contours of the world economy* (5th ed.). Guilford Press.

Douglass, M. (1988). The transnationalization of urbanization in Japan. *International Journal of Urban and Regional Research, 12*(3), 425–454. https://doi.org/10.1111/j.1468-2427.1988.tb00088.x

Douglass, M. (1993). The "new" Tokyo story: Restructuring space and the struggle for place in a world city. *Japanese Cities in the WORLD Economy*, 83–119. https://ci.nii.ac.jp/naid/10003849086/en/

Douglass, M. (2000). Mega-urban regions and world city formation: Globalisation, the economic crisis and urban policy issues in Pacific Asia. *Urban Studies, 37*(12), 2315–2335.

Dufoix, S. (2008). *Diasporas*. University of California Press.

Duplan, K., & Cranston, S. (2023). Towards geographies of privileged migration: An intersectional perspective. *Progress in Human Geography, 47*(2), 333–347. https://doi.org/10.1177/03091325231156927

Fainstein, S. S. (2001). Inequality in global city-regions. *disP - ThePlanning Review, 37*(144), 20–25.

Fainstein, S. S. (2011). Inequality in global city-regions. In A. J. Scott (Ed.), *Global city-regions: Trends, theory, policy*. Oxford University Press.

Faist, T. (2000). *The volume and dynamics of international migration and transnational social spaces*. Oxford University Press.

Faist, T. (2010). Towards transnational studies: World theories, transnationalisation and changing institutions. *Journal of Ethnic and Migration Studies, 36*(10), 1665–1687.

Faulconbridge, J. R. (2007). Relational networks of knowledge production in transnational law firms. *Geoforum, 38*(5), 925–940.

Faulconbridge, J. R. (2008). Managing the transnational law firm: A relational analysis of professional systems, embedded actors, and time—space-sensitive governance. *Economic Geography, 84*(2), 185–210.

Friedmann, J. (1986). The world city hypothesis. *Development and Change, 17*(1), 69–83.

Friedmann, J., & Wolff, G. (1982). World city formation: An agenda for research and action. *International Journal of Urban and Regional Research, 6*(3), 309–344.

Fujita, K. (1991). A world city and flexible specialization: Restructuring of the Tokyo metropolis. *International Journal of Urban and Regional Research, 15*(2), 269–284. https://doi.org/10.1111/j.1468-2427.1991.tb00634.x

GaWC. (2020). *The World according to GaWC.* Globalization and world cities research network. Retrieved September 15, 2023, from https://www.lboro.ac.uk/microsites/geography/gawc/gawcworlds.html

Glick-Schiller, N. (2012). Transnationality, migrants and cities: A comparative approach. In A. Amelina, D. D. Nergiz, T. Faist, & N. Glick Schiller (Eds.), *Beyond methodological nationalism* (pp. 23–40). Routledge.

Glick-Schiller, N., Basch, L., & Blanc-Szanton, C. (1992). Towards a definition of transnationalism. *Annals of the New York Academy of Sciences, 645*(1).

Goulbourne, H., Reynolds, T., Solomos, J., & Zontini, E. (2010). *Transnational families: Ethnicities, identities and social capital.* Routledge.

Greiner, C., & Sakdapolrak, P. (2013). Translocality: Concepts, applications and emerging research perspectives. *Geography Compass, 7*(5), 373–384.

Guarnizo, L. E. (1997). The emergence of a transnational social formation and the mirage of return migration among Dominican transmigrants. *Identities, 4*(2), 281–322. https://doi.org/10.1080/1070289X.1997.9962591

Guarnizo, L. E., Portes, A., & Haller, W. (2003). Assimilation and transnationalism: Determinants of transnational political action among contemporary migrants. *American Journal of Sociology, 108*(6), 1211–1248.

Hall, P. (1966). *The world cities.* McGraw-Hill.

Hamnett, C. (1994). Social polarisation in global cities: theory and evidence. *Urban Studies, 31*(3), 401–424.

Haug, S. (2000). *Klassische und neuere Theorien der Migration.* Mannheimer Zentrum für Europäische Sozialforschung.

Hayes, M., & Zaban, H. (2020). Transnational gentrification: The crossroads of transnational mobility and urban research. *Urban Studies, 57*(15), 3009–3024.

Heil, T. (2020). *Comparing conviviality: Living with difference in Casamance and Catalonia*. Springer Nature.
Hill, R. C., & Kim, J. W. (2000). Global Cities and Developmental States: New York, Tokyo and Seoul. *Urban Studies, 37*(12), 2167–2195. https://doi.org/10.1080/00420980020002760
Johnson, C. (1982). *MITI and the Japanese miracle, the growth of industrial policy, 1925–1975*. Stanford University Press.
Kamo, T. (1988). *Toshi no Seijigaku [Politics of the City]*. Jichitai Kenkyusha.
Kamo, T. (2000). An aftermath of globalisation?: East Asian economic turmoil and Japanese cities adrift. *Urban Studies, 37*(12), 2145–2165. https://doi.org/10.1080/00420980020002751
Kamo, T. (2005). *Sekai toshi: Toshi saisei no jidai no naka de*. Yūhikaku.
Kearney, M. (1995). The local and the global: The anthropology of globalization and transnationalism. *Annual Review of Anthropology, 24*(1), 547–565.
Khagram, S., & Levitt, P. (2008). *The transnational studies reader. Intersections and innovations*. Routledge.
King, R. (2012). Geography and migration studies: Retrospect and prospect. *Population, Space and Place, 18*(2), 134–153.
Kivisto, P. (2001). Theorizing transnational immigration: A critical review of current efforts. *Ethnic and Racial Studies, 24*(4), 549–577.
Kloosterman, R., Van Der Leun, J., & Rath, J. (1999). mixed embeddedness: (in)formal economic activities and immigrant businesses in The Netherlands. *International Journal of Urban and Regional Research, 23*(2), 252–266.
Knox, P. L., & Taylor, P. J. (1995). *World cities in a world-system*. Cambridge University Press.
Krätke, S., Wildner, K., & Lanz, S. (2012). *Transnationalism and urbanism* (Vol. 25). Routledge.
Kunz, S. (2016). Privileged mobilities: Locating the expatriate in migration scholarship. *Geography Compass, 10*(3), 89–101.
Lees, L., Shin, H. B., & López-Morales, E. (2016). *Planetary Gentrification*. Polity Press.
Levitt, P. (2001). *The transnational villagers*. University of California Press.
Levitt, P., DeWind, J., & Vertovec, S. (2003). International perspectives on transnational migration: An introduction. *International Migration Review, 37*(3), 565–575.
Machimura, T. (1992). The urban restructuring process in Tokyo in the 1980s: Transforming Tokyo into a world city. *International Journal of Urban and*

Regional Research, 16(1), 114–128. https://doi.org/10.1111/j.1468-2427.1992.tb00468.x

Machimura, T. (1994). *"Sekai toshi" Tōkyō no kōzō tenkan: Toshi risutorakuchuaringu no shakaigaku* (Shohan ed.). Tōkyō Daigaku Shuppankai.

Machimura, T. (1998). Symbolic use of globalization in urban politics in Tokyo. *International Journal of Urban and Regional Research, 22*(2), 183–194. https://doi.org/10.1111/1468-2427.00134

Marcuse, P. (2000). Cities in quarters. In G. Bridge & S. Watson (Eds.), *A companion to the city* (pp. 270–281). Blackwell Publishing.

Massey, D., Arango, A., Hugo, G., Koudouchi, A., Pellegrino, A., & Taylor, J. E. (2005). *Worlds in motion. Understanding international migration at the end of the millennium.* Oxford University Press.

Meissner, F., & Vertovec, S. (2015). Comparing super-diversity. *Ethnic and Racial Studies, 38*(4), 541–555. https://doi.org/10.1080/01419870.2015.980295

Morgan, G. (2001). Transnational communities and business systems. *Global Networks, 1*(2), 113–130.

Ong, A. (1999). *Flexible citizenship: The cultural logics of transnationality.* Duke University Press.

Ossenbrügge, J. (2005). Transstaatliche, plurilokale und glokale soziale Räume - Grundbegriffe zur Untersuchung transnationaler Beziehungen und Praktiken. In J. Ossenbrügge & M. Reh (Eds.), *Social spaces of African societies* (pp. 15–34). Lit Global [distributor].

Parnreiter, C. (2012). Conceptualizing transnational urban spaces: Multicentered agency, placeless organizational logics, and the built environment. In S. Krätke, K. Wildner, & S. Lanz (Eds.), *Transnationalism and urbanism* (pp. 91–111). Routledge.

Pollock, D. C., & Van Reken, R. E. (2001). *The third culture kid experience: Growing up among worlds.* Intercultural Press.

Portes, A. (2001). Introduction: The debates and significance of immigrant transnationalism. *Global Networks-a Journal of Transnational Affairs, 1*(3), 181–194.

Portes, A., Guarnizo, L. E., & Landolt, P. (1999). The study of transnationalism: Pitfalls and promise of an emergent research field. *Ethnic and Racial Studies, 22*(2), 217–237.

Portes, A., Haller, W. J., & Guarnizo, L. E. (2002). Transnational entrepreneurs: An alternative form of immigrant economic adaptation. *American Sociological Review, 67*(2), 278–298.

Pries, L. (1997). *Transnationale migration* 1. Aufl. Nomos.
Pries, L. (1999). *Migration and transnational social spaces*. Ashgate.
Pries, L. (2001). *New transnational social spaces. International migration and transnational companies in the early twenty-first century*. Routledge.
Pries, L. (2005). Configurations of geographic and societal spaces: A sociological proposal between 'methodological nationalism' and the 'spaces of flows'. *Global Networks, 5*(2), 167–190.
Reckwitz, A. (2002). Toward a theory of social practices: A development in culturalist theorizing. *European Journal of Social Theory, 5*(2), 243–263.
Rimmer, P. J. (1986). Japan's world cities: Tokyo, Osaka, Nagoya or Tokaido Megalopolis? *Development and Change, 17*(1), 121–157. https://doi.org/10.1111/j.1467-7660.1986.tb00233.x
Roberts, B. R., Frank, R., & Lozano-Ascencio, F. (1999). Transnational migrant communities and Mexican migration to the US. *Ethnic and Racial Studies, 22*(2), 238–266.
Saito, A. (2003). Global city formation in a capitalist developmental state: Tokyo and the waterfront sub-centre project. *Urban Studies, 40*(2), 283–308. https://doi.org/10.1080/00420980220080281
Saito, A., & Thornley, A. (2003). Shifts in Tokyo's world city status and the urban planning response. *Urban Studies, 40*(4), 665–685. https://doi.org/10.1080/0042098032000065245
Sassen, S. (1991). *The global city: New York, London, Tokyo*. Princeton University Press.
Sassen, S. (1996). Whose city is it? Globalization and the formation of new claims. *Public Culture, 8*(2), 205–223.
Sassen, S. (2011). *Cities in a world economy*. Sage Publications.
Saxenian, A. (2002). *Local and global networks of immigrant professionals in Silicon Valley*. Public Policy Institute of California.
Saxenian, A. (2007). *The new argonauts: Regional advantage in a global economy*. Harvard University Press.
Schapendonk, J. (2021). Counter moves. Destabilizing the grand narrative of onward migration and secondary movements in Europe. *International Migration, 59*(6), 45–58. https://doi.org/10.1111/imig.12923
Schatzki, T. R. (1996). *Social practices: A Wittgensteinian approach to human activity and the social*. Cambridge University Press.
Schatzki, T. R., Knorr-Cetina, K., & Von Savigny, E. (2001). The practice turn in contemporary theory 44. Routledge.

Sigler, T., & Wachsmuth, D. (2016). Transnational gentrification: Globalisation and neighbourhood change in Panama's Casco Antiguo. *Urban Studies, 53*(4), 705–722.

Sklair, L. (2001). *Transnational capitalist class*. Blackwell.

Smith, M. P. (2001). *Transnational urbanism: Locating globalization* (Vol. 221). Blackwell Publishers Malden.

Smith, M. P. (2005). Transnational urbanism revisited. *Journal of Ethnic and Migration Studies, 31*(2), 235–244. https://doi.org/10.1080/1369183042000339909

Smith, M. P., & Guarnizo, L. (Eds.). (1998). *Transnationalism from below*. Transaction Publishers.

Taylor, P. (2004). *World City Network: A Global Urban Analysis*. Routledge

Taylor, P. J., Ni, P., Derudder, B., Hoyler, M., Huang, J., & Witlox, F. (2011). *Global urban analysis: A survey of cities in globalization*. Routledge.

Valentine, G. (2008). Living with difference: Reflections on geographies of encounters. *Progress in Human Geography, 32*(3), 323–337.

Vertovec, S. (1999). Conceiving and researching transnationalism. *Ethnic and Racial Studies, 22*(2), 447–462.

Vertovec, S. (2007). Super-diversity and its implications. *Ethnic and Racial Studies, 30*(6), 1024–1054.

Vertovec, S. (2009). *Transnationalism*. Routledge.

Vincent, C., Neal, S., & Iqbal, H. (2017). Encounters with diversity: Children's friendships and parental responses. *Urban Studies, 54*(8), 1974–1989.

Voigt-Graf, C. (2004). Towards a geography of transnational spaces: Indian transnational communities in Australia. *Global Networks-a Journal of Transnational Affairs, 4*(1), 25–49.

Waters, J. (2003). Flexible citizens? Transnationalism and citizenship amongst economic immigrants in Vancouver. *The Canadian Geographer/Le Gèographe canadien, 47*(3), 219–234.

Wehrhahn, R., & Tölle, A. (2016). *Translokalität und lokale Raumproduktionen in transnationaler Perspektive*. Logos Verlag.

Wessendorf, S. (2013). Commonplace diversity and the 'ethos of mixing': Perceptions of difference in a London neighbourhood. *Identities, 20*(4), 407–422.

Wessendorf, S. (2014). 'Being open, but sometimes closed': Conviviality in a super-diverse London neighbourhood. *European Journal of Cultural Studies, 17*(4), 392–405.

White, J. W. (1998). Half-empty bottle or no bottle at all?: A rejoinder to Sassen and Smith. *Urban Affairs Review, 33*(4), 489–491. https://doi.org/10.1177/107808749803300404

Wilson, H. F. (2017). On geography and encounter: Bodies, borders, and difference. *Progress in Human Geography, 41*(4), 451–471. https://doi.org/10.1177/0309132516645958

Wimmer, A., & Glick Schiller, N. (2002). Methodological nationalism and beyond: Nation–state building, migration and the social sciences. *Global Networks, 2*(4), 301–334.

Yamamura, S. (2022a). Transnational migrants and the socio-spatial superdiversification of the global city Tokyo. *Urban Studies, 59*(16), 3382–3403.

Yamamura, S. (2022b). Transnationalism, elite migrants, and spatiality of superdiversity. In F. Meissner, N. Sigona, & S. Vertovec (Eds.), *The Oxford Handbook of Superdiversity*. Oxford University Press. https://doi.org/10.1093/oxfordhb/9780197544938.013.18

Yamamura, S. (2023). The multi-scalar embeddedness of support policies for migrant entrepreneurship in Japan. *International Migration, 61*(2), 67–86.

Yamamura, S., & Lassalle, P. (2020). Approximating entrepreneurial superdiversity: Reconceptualizing the superdiversity debate in ethnic minority entrepreneurship. *Journal of Ethnic and Migration Studies, 46*(11), 2218–2239.

Yamamura, S., & Lassalle, P. (2022). Extending mixed embeddedness to a multi-dimensional concept of transnational entrepreneurship. *Comparative Migration Studies, 10*(14). https://doi.org/10.1186/s40878-022-00288-y

Yamamura, S., Lassalle, P., & Shaw, E. (2022). Intersecting where? The multi-scalar contextual embeddedness of intersectional entrepreneurs. *Entrepreneurship & Regional Development, 34*(9–10), 828–851. https://doi.org/10.1080/08985626.2022.2120086

Ye, J. (2017). Managing urban diversity through differential inclusion in Singapore. *Environment and Planning D: Society and Space, 35*(6), 1033–1052. https://doi.org/10.1177/0263775817717988

Yeoh, B. S. (2006). Bifurcated labour: The unequal incorporation of transmigrants in Singapore. *Tijdschrift voor Economische en Sociale Geografie, 97*(1), 26–37.

3

Putting People's Spaces into Place

The spatial dimension of transnational migration as well as transnational urbanism has been somewhat neglected in research, despite the actual ubiquitous spatial turn that occurred in cultural and social sciences and also despite the extensive debate on spatiality within urban studies, in particular set forth by urban geographers and sociologists. The concept of superdiversity mentions socio-spatial diversification and also explicitly points out the importance of the urban lens for the unit of analysis, yet a spatial conceptualization of the urban phenomena still lacks. This chapter therefore briefly looks into the debates on spatialities, but specifically in the context of the three pillars of the here present study, that is, the city (in particular global cities), the migration (in particular transnational migration) and the space (in particular transnational space) as these fields are those under scrutiny. It first gives an overview on the theoretical developments within and around geography. Then, by pointing out the different contrasting strands of the approaches, it shows the consequential gap on the research field of migrants *and* cities. And, then it presents how the study of the transnational space itself, in which people's (transnational migrants) spaces are put into place, can fill this gap.

S. Yamamura, *Spatial Diversity in the Global City*, Global Diversities,
https://doi.org/10.1007/978-3-031-64725-3_3

In its overall intention and function for the presented work, this chapter presents the theoretical background on the already discussed and identified gap on socio-spatial diversification in cities and proposes the methodological approach of the empirical work. The methodological approach derives from the theoretical take on the actual issue on the spatial diversification. The revival of methods such as the mental map and the combination of different qualitative methods within an overall mixed-methods approach to the research field is well reasoned for the given exploratory research and a novelty for the under-researched field. However, it should be noted that further methodological innovation, especially incorporating digital methods, could give further nuances and interesting insights into the complexity of the spatiality of urban superdiversity. In fact, as it will be discussed later on, the importance of the digital connectedness and, thus, the emergence of hybrid and digital spaces as part of the transnationally highly mobile migrants' everyday socio-spatial practices require methodological innovation in the fieldworks following up to this work.

3.1 Debates on Spatialities in the Urban Context

Social sciences, including human geography, have been experiencing different turns over the academic history. The most recent being, as explained previously, the transnational turn within migration research, but also one of the main turns in the last two decades is the spatial turn in cultural and social sciences. The spatial turn, as a matter of fact, has even spilled over to adjacent disciplines, such as to economic geography or business studies.

An interesting trend regarding this spatial turn, though, is that geography as a discipline, in particular human geography, is by definition a spatially focused discipline, where it appears paradoxical for it to be experiencing a spatial turn within itself. In fact, the spatial turn in geographical science had occurred with a precursory time shift, already in the 1960s. Such spatial turn was most evident in Germany, where geography with its strong roots in geopolitics and its development into an area and regional science (somewhat in the postcolonial path-dependency), turned its back to a more holistic understanding of human geography occurred in 1969 at the *Kieler Geographentag*. It is by no coincidence that this was in times of the post-war student movements and the generation's emancipation from the past. This caesura encompassed, on one side, the shift to a new understanding of the spatial away from the space as a container and also from a quantitative approach. On the other, the incorporation of and turn to a more social scientific understanding of spatially related issues was addressed. Rooting in the critique of theoretical deficits of regional geography, the spatial turn in geography blossomed in the new development of social geography. Whereas the social-science oriented geography in the Anglo-Saxon academia diversified subsequently into different strands in the 1970s, also merging with cultural scientific approaches, the German geography caught up with the trend with a slight time lag in the 1980s. On the theoretical agenda of human geography, critical strands such as Marxist, feminist, post-colonialist, post-modernist, structuralist or actor-based perspectives established. And yet again, critical of structuralist approaches of Marxist ideas, further debate flourished especially reflecting societal movements in urban and spatial contexts, where especially Marxist and Neo-Marxist scholars like Lefebvre and Bourdieu came into focus, and one recent strand within geography turned further to discussing structurational approaches of Giddens.

Basing on these different strands within geography, which also coincide with larger social theoretical debates, the understanding of spatial issues, including that on the urban context, profoundly differs within human geography. Whereas the cultural turn in the 1980s caused the strand of the new cultural geography to flourish, bringing the idea of attribution of meanings to space, thus creating places, and by that differentiating the notion of *space* from *place*, but also breaking with the geodeterminist "old" cultural geography, the relational turn put forward the importance of interdependencies of different geographies of scales. Such relational turn in human geography is also reflected in the economic geography, where the intertwining of "socio-spatial relations of economic actors (…) with processes of economic change at various geographical scales" (Yeung, 2005: 37) is being researched. In close proximity to the economic geographical debate, the relational turn is also partly observable in urban studies, where global city researchers at the intersection of economic and urban geography focus on the embeddedness and interconnectedness of cities within a global network.

3.2 Transnational Space(s) and the Making

The transnational space in literature is somewhat a blurry terminology. It is not only that "the distinction between countries of origin and destination blurs" (Faist, 2000: 13) in the transnational migration context. In some cases, the transnational social space is even apprehended as a metaphoric construct:

"The metaphor of transnational social spaces helps to broaden the scope of migration studies to include the circulation of ideas, symbols, and material culture, not only the movement of people" (Goldring, 1996: 77), implying the intangible nature of the space as a construct.

Although agreeing with the basic assumption of the transnational space being a scientific concept and construct, spatial phenomena though, including the transnational spaces, are tangible in the real world. Social practices and interactions are located in specific spaces so are transnational spaces. There is indeed a need to further develop and concretize the idea of the diffuse and intangible terms of transnational spaces, such as the "deterritorialized spaces" that emerge "above and beyond" (cp. Glick-Schiller et al., 1992, 1995) concrete territorial spaces.

In fact, transnational spaces in migration research have been primarily discussed as transnational *social* spaces. With strong roots of migration research, particularly transnational migration research, in anthropology and sociology, spaces have been regarded as social spaces in the Bourdieusian sense of social fields or habitus. This main approach of sociology is dominant in the debates on transnational spaces. Two prominent sociologists have extensively researched and theorized on transnational social space, that is, Ludger Pries (2001a) and Thomas Faist (2000).

Transnational spaces are, according to these sociologists, defined as a "combination of sustained social and symbolic ties, their contents, positions in networks and organizations, and networks of organizations that can be found in multiple states" (Faist, 2000: 199). Faist further distinguishes three forms of transnational social spaces, which are "transnational kinship groups: reciprocity," "transnational circuits: constant circulation of goods, people and information" and "transnational communities: connected by dense and strong social and symbolic ties over time and across space" (ibid.: 202). On a more social theoretical side, Pries defines it as "dense, stable, pluri-local and institutionalized frameworks composed of material artifacts, the social practices of everyday life, as well as systems of symbolic representation that are structured by and structure human life" (Pries, 2001a: 5).

Faist acknowledges that space from the perspective of social geography does not necessarily refer only to physical features, but also encompasses "a social meaning that extends beyond simple territoriality," thus, is distinct from the notion of place, and also "only with concrete social or symbolic ties does it gain meaning for potential migrants" (Faist, 2000: 46). However, taking social theories into account, Faist comes to the conclusion that the geographical aspect of social space becomes superfluous in the transnational social space, respectively that "the process of separation of space from place is essential to modernization" (ibid.: 197). His assumption appears to be correct that "physical location and geographical distance are not the only grid upon which migrants map collective political action, cultural codes, and economic cooperation" (ibid.). And Pries presents a similar idea of the disruption of social and geographic space," pointing out that "a social space can (…) expand over several and distinct geographical spaces" and there are "social spaces that have a multi-local geographic link rather than an exclusive one" (Pries, 2001b: 57).

These assumptions, however, draw the question whether geographical space itself can be distinguished so clearly from the social space at all, respectively what kind of underlying spatial concept is being taken into consideration. The actual point is that geography, in particular cultural geography, has evolved further since the spatial and then cultural turn, and incorporates social as well as geographical aspects as part of their definition of spaces and places. However, as King notes (2012), geographers have only recently taken up the concept of transnationalism and are rather new in the field of migration research.

One exception within the otherwise sociological debate on transnational space is the edited publication by Jackson, Craig and Dwyer on the "spaces of transnationality" (2004). Shortly after Pries and Faist published their works on the transnational space, they appear to be voicing their geographer's perspective, pointing out that "[r]esearch on transnational space is still in its infancy. By outlining some of the complex geographies (…), we are aware of how much yet remains to be done" (Jackson et al., 2004, S. 1), but also criticizing the lack of geography in the transnationalism debate:

> studies of the economic, political and cultural dimension of transnationalism have characteristically under-played the transformation of space that is involved in the evolution of transnational social forms. Rather than taking space as a passive backdrop to transnational social relations, we argue that space is constitutive of transnationality in all its different forms. (Jackson et al., 2004: 7)

Listing up four contexts of transnationalism which emphasizes the significance of space, they include the fact that transnationality has been developed in "different geographical spaces," yet varying over time and space, which can in turn be distinguished by their "particular spatialities," then mention the "cartographies" of migration and finally argue that:

> transnationality is a geographical term, centrally concerned with reconfigurations in relations with place, landscape and space. (…) So, the national is about place, territory, landscape rootedness, belonging; the transnational connotes space, de-territorialization, other global cultural '-scapes' like mediascapes (Appadurai, 1996), uprooting, rootlessness and routes of travel, and exclusion and longing. (Jackson et al., 2004, S. 1–4)

With their broad definition which aligns with the recent understanding of space in cultural and social geography, that is:

> Our use of the term encompasses all of those engaged in transnational cultures, whether as producers or consumers. It includes not just the material geographies of labour migration or the trading in transnational goods and services but also the symbolic and imaginary geographies through which we attempt to make sense of our increasingly transnational world. (Jackson et al., 2004: 2ff.)

The transnational space in the end does not seem to get more tangible than previous concepts.

Also, with regard to the contributions in their edited publication, it is questionable if the actual spatial extent of transnationalism is captured. Contributions cover indeed a different approach to the otherwise often bidirectional transnationalism, such as the idea that a single community is dispersed over multiple location. Yet, what follows are discussions on transnational commodity space, transnational commodity culture in the lived social field, and also more traditional accounts of transnationalism with transnational social fields maintained by transnational links (through remittances, frequent travels and interactions), or state-sponsored transnational migration policies (Jackson et al., 2004). While these are highly relevant topics of transnationalism, it remains yet once again blurry how exactly the transnationalism is reflected in the actual geographical sense.

One important take away from the geographers' take on transnationalism is their statement the focus on transnational spaces instead of merely identifying transnational communities "opens up ways of exploring this multiplicity of transnational experiences and relations" (ibid.: 3). However, also being aware of the "complex, multidimensional and multiply inhabited" (Crang et al., 2004) nature of transnational space, there is actually a need to find an analytical tool to make these multidimensional spaces tangible.

Instead of going into the details of different spatial theoretical approaches within geography and social sciences, and discussing these extensively, what is proposed here for the focus on empirically capturing the socio-spatial diversification in the global city is an analytical framework which enables the researcher to empirically grasp spatial phenomena, that of Martina Löw. As intellectually intriguing and inspiring the discussion of different spatial theories may be, for the purpose of the empirical investigation into the socio-spatial diversity produced by transnational migrant professionals, it is crucial to have an operationalizable theoretical take. While this monograph acknowledges the meaning and impact of the structures of global economy and macro-level policies in which transnational migrants are embedded, the actual agency of transnational migrants in the production of space including their contribution to socio-spatial diversification in the global cities in which they act need to be considered, too. By bringing an actor-theoretical approach and theorizing space from a structurational approach based on Giddens, Löw's spatial theory bring the balance between the structure and agency which characterizes the transnational space-making.

The space is understood as a "relational order/arrangement of social goods and living beings in places" (Löw, 2001: 224; author's translation). The usefulness of this approach particularly then lies on the analytical framework. The space is constituted by two analytically differing processes,[1] that is, the so-called spacing and the synthesis performances. The spacing process encompasses the allocation of physical materials into space, such as arrangements of architectural buildings and other objects, but also the placing of people and symbols into space. Through the latter synthesis performance such goods and people are combined to spaces through "imaginative, perceptive, and reminiscential processes" (ibid: 225, author's translation).[2]

This dual analytical framework signifies that any person and any social group construct specific spaces on the basis of their specific experience and perception. Individual spaces then through collective experiences, behaviors and other perceptive processes layer together to a collective space. In the case of transnational professionals, it can be objectively stated that they show specific habits and actions as part of the transnational corporate context, where they are unified in their transnational "new management mentality" (Bartlett & Ghoshal, 2002: 20) which are imposed on by the firms. This side would be determined by their careers as transnational professionals. At the same time, through their active choice to take up such a mobile transnational life, thus, having clear agency to mobility, there is a likelihood that they perceive, imagine and

---

[1] The analytical framework is dual in nature but this does not necessarily equate with the duality of the space. As a matter of fact, the space should be understood as being multiple in identities and also dynamics according to human geographical concepts as Doreen Massey states.

[2] The author is aware that the terminology and the conceptual understanding of space and place differ from the cultural geographical tradition (with places being spaces with attached meanings, see e.g. Peter Jackson, 1989). However, as an empirical analysis of the place-making in case of transnational professionals are in fact pragmatically very challenging and the monograph is embedded in the overall socio-spatial patterns not only within a specific city but also on the larger framework of global cities networks, respectively the globe, the analytical framework by Löw provides potentials for a new approach for empirical research on transnational spaces.

remember their environment on the basis of their experience and associations in a similar manner, whereas this part is to be empirically researched. The reflection of the presupposed transnationality through their profession on their private socio-spatial patterns and their cognitive processes is the object of research. The extent and intensity of transnational social practices apart from their professional daily lives can offer the clues to the overall transnational space. At the same time, if transnational professionals describe specific spatial aspects of the places in which they interact with other people and in which they work and live, then it will also be possible to draw conclusions from their statements on the concrete physical and, thus, tangible features of the transnational space, too.

Although this spatial theory might at first sight convey the idea that the transnational space is solely constructed by the individual through the spacing and synthesis performances, it must be noted that this is not necessarily the case. The spatial theory of Löw derives from structurational ideas of, inter alia, Giddens, taking both structures and agencies into consideration. The embeddedness of the transnational professionals' choices over socio-spatial patterns will be discussed in the empirical chapters and later also developed in its theoretical context discussing the balance of the structure and agency.

3.3 Methodological Implications

The methodological implication of this dual analytical approach is that both sides of the synthesis and spacing processes should but also can be conceptually separated. The presented work follows an embedded mixed-methods design (in the sense of Creswell & Clark, 2007; Creswell et al., 2003), whereas quantitative data are mainly used for the overall context of grasping the global city phenomenon concerning ranking and connectivities calculated by the Globalization and World Cities (GaWC) researchers. Further secondary data were consulted to grasp the city- and regional-level residential and migrant population as background knowledge for the actual qualitative research. As it can be seen in the visualization of the research design below (Fig. 3.1), the actual empirical qualitative research consists of different methods:

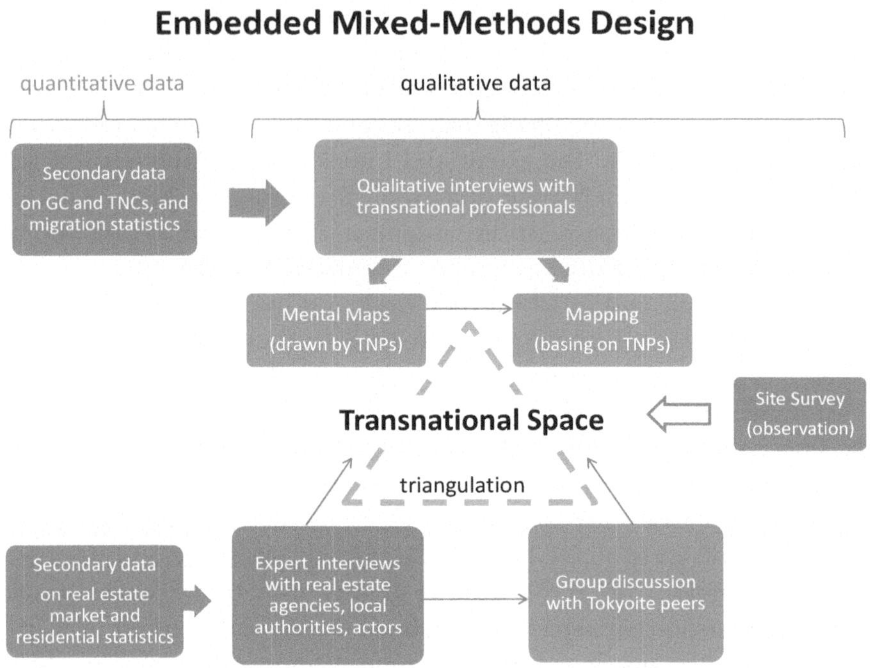

Fig. 3.1 Triangulation of transnational space by embedded mixed-methods design (author's own)

Problem-Oriented Qualitative Interviews

The main source of information comes from qualitative interviews, specifically problem-oriented interviews (cf. Witzel & Reiter, 2012), with transnational professionals. In total 45 persons of transnational corporations in the financial industry (NB: only two female, yet this reflects the gender distribution in the occupation at the organizational hierarchical level) were found through snowball technique. These transmigrants were chosen according to four main criteria to ensure consistency within the group, thus, allow intragroup comparability:

1. Financial professional of a transnational corporation to accommodate a methodologically unified specific social group but also with

comparable occupational contexts embedded in the global city theory (FIRE sector);

2. Managerial position as only those at the higher level can be regarded as part of those who are transnational according to economic geographical definition and theoretically also having agency within the corporate context (being directly involved in decision-making and the processes of transnationalization businesswise);
3. Extensive experiences abroad, both long and/or short-term assignments overseas so to have transnational migrants who show daily social practices crossing national borders and also having moved themselves;
4. Work and residential experience in the global city Tokyo; as the fluctuation, respectively mobility among financial professionals are very high, and the access to the field very limited, those who have moved to another global city have also been taken into account;
5. One aspect that has been consciously ignored for the selection of these transmigrants was the ethnic or national backgrounds. The collective membership to the transnational professionals group should be the key to the new empirical approach of a transnational migration research breaking with the ethno-focal lens. Nationalities and ethnicities that were found were accordingly diverse: US-American, UK British including Scottish and English, Australian, German, Japanese, Indian, Singaporean, French, Italian, Korean, Chinese-Singaporean, Slovak (including mixed-backgrounds).

Interviews lasted from 30 minutes up to 2.5 hours (on average 45–60 minutes), yet the time constraint was unavoidable due to the occupational context of transnational financial professionals (cf. Useem, 1995 on interviewing corporate executives). Similar constraints were the usage of also telephone-interview due to the high mobility. As common in executive and elite research, the access to the field was limited so that gatekeepers and snowballing-technique was used to reach professionals with the specific characteristics.

The "problem" of the problem-oriented interviews was the socio-spatial patterns of the transnational professionals as migrants. After a first introductory part on their individual mobility and migration background, that is, experiences abroad but also current frequency of business trips, the first part focused on the daily practices in the transnational corporation and its business community/networks. The main part then focused on their socio-spatial patterns stretching from formal, over semi-formal to informal and private social interactions. It was explicitly discussed with which persons groups they interacted where for which purposes. These socio-spatial patterns were first focused on Tokyo as an example and then developed more freely to general patterns at travels, both business and private, as well as in other global cities they had been to before. With reflective discussions on their socio-spatial patterns also encompassing changes in their life courses, then phased out to issues of reflections on their cultural or professional identities as well as their perception and evaluation of their own lifestyles.

The interviews were conducted mainly in English, but where requested or naturally occurring Japanese or German were also spoken. In some cases, interviewees code-switched during the interview or even within sentences. The interviews were recorded and transcribed for further analysis. Data were selectively coded according to the approach of qualitative content analysis by Mayring (1994).

Mental Mapping

A novel, respectively revived, approach to complement these qualitative interviews was the usage of mental maps. Mental maps (Downs & Stea, 2005; Lynch, 1960; White & Gould, 1986) were developed in the 1960s as part of perceptive geography. They are here, however, used to reflect the synthesis performances of the spatial analytical framework. It bases on the premises that maps drawn by individuals reflect both the knowledge and personal perception of the environment. In accordance with studies using mental maps, it could be expected that these individuals of the similar professional background draw similar patterns into maps, too. And differences in socio-spatial patterns within the group would also be reflected in them. Being explicitly asked for an ego-centric map of Tokyo as well as of their own world, in which they live in and having special regions, cities or areas indicated that were important for them in both business and private contexts, the transnational professionals reproduced the analytical process suggested in the spatial concept. The concrete analytical processes and their implications on the micro (Tokyo) and the macro-level (world map) will be discussed further in each subsection of the empirical chapters (Chaps. 4–7).

Mapping and Site Survey

As a further means to empirical grasp the transnational spaces presented by the interviewees, locational maps were drawn on the basis of the information from the interviews, and site surveys carried out. Ethnographic assessment (Hall, 2015) of the urban landscape, including hints on the linguistic landscape as well as observances of the different people in these areas, complemented the interviews.

Triangulation of the Transnational Space

To triangulate the information deriving directly from the transnational migrants, two sets of further qualitative interviews were conducted. On the one hand, to better understand the choices on socio-spatial patterns, including the residential choices, of transnational professionals, five expert interviews (using the term in the sense of Meuser & Nagel, 1991) with the major real estate agencies specialized in housing rentals for highly skilled corporate foreigners and Japanese professionals were conducted. Moreover, officers of the metropolitan administration in the Minato ward (the district within Tokyo where all professionals were residing and showing their socio-spatial patterns located) were interviewed regarding policies on the foreign population. On the other hand, after different patterns emerged within the transnational professionals, with one more strongly appearing to be attached to the local upper class, a focus group discussion with Japanese Tokyoites were also conducted. These qualitative interviews triangulated the transnational professionals' perspective on the city of Tokyo and their views on the structures and nature of urban embeddedness.

3.4 Empirical Research Process

Against the background of the a priori concepts that underlie the overall research context, which is the global cities phenomena, the transnational migration as well as the proposed spatial concept, the research follows an abductive research process (see visualization of the research process, Fig. 3.2).

Framed in these a priori concepts (1), qualitative interviews with the primary subjects, that is, "transmigrants from above," are first interviewed (2). Following an inductive analysis of these qualitative data (3), socio-spatial patterns are filtered and empirics, including "surprising research evidence" are inductively analyzed according to the ideas of the analytical framework on the production of transnational space (4). This analysis is then qualitatively triangulated by the aforementioned expert interviews and focus group discussions, which in turn abductively confirmed the empirical evidence (5). Emerging conceptualizations of socio-spatial

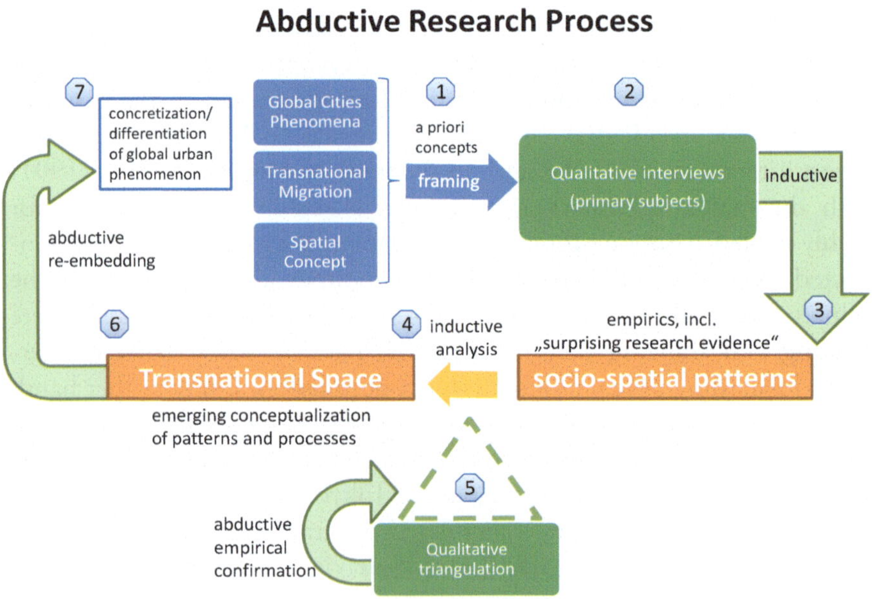

Fig. 3.2 Abductive research process on transnational spaces (author's own)

patterns and processes clarified the transnational spaces (6), which then were abductively re-embedded into the original concepts (7). In this last step, as empirical evidence had emerged the concepts of the spatial diversification in global cities could be concretized and more clearly differentiated as a global urban phenomenon.

As it can be deduced from this rather dry theoretical and methodological chapter, the capturing of transnational space if taken more literary in the geographical sense and not accepting the sociological and symbolic term can become quite complex. The embedded mixed-methods design as presented above is an attempt to operationalize the spatial theoretically underlying idea of transnational space. While the combination of different methods and the abductive research process seem to allow a more holistic analysis of the transnational space and then of the overall socio-spatial diversification in global cities, it is by no means the one and only approach that is possible for the fieldwork. In fact, as it will be discussed in the conclusive chapter, there are even more aspects that could be further investigated with other means and—as it is usually the case for social scientific work—with more time and capacities.

What is important to take from this chapter is the actual theoretical grounding of the methodological approach, respectively the coherence of the analytical framework of the introduced spatial theory with the actual topic. The transnational space, as discussed in the previous chapters in the different overviews of the scientific debates, is an under-researched and somewhat over-passed topic because of the different disciplinary approaches to transnationalism. At the same time, the vacuum surrounding transnational professionals' spaces also reflects the theoretical approaches to focus on either the structures (economic geographical and socio-political scientific views on transnationalism from above) or the agency of transnational migrants (anthropological and sociological takes on transnationalism from below). For closing the gap on transnational spaces, and later further discussing the socio-spatial diversification mechanism of urban spaces, a more balanced spatial theoretical approach that accommodates both the structures of mobilities as well as the agency of the mobile is essential.

Starting with a more historical take on the development of Tokyo to a global city, the following chapters will now present the empirical results of the study. The empirics will be introduced in chapters that can be read separately, too, as each of them is an enclosed story from different perspectives. The first three empirical Chaps. (4–6) present different narratives and patterns of transnational spaces within Tokyo, thus taking an urban-scale lens. These will then in the subsequent parts be discussed from a global city-regional, thus, regional scale, and finally from larger context of global-scale networks of global cities.

References

Appadurai, A. (1996). *Modernity at large: Cultural dimensions of globalization*. University of Minnesota Press.

Bartlett, C. A., & Ghoshal, S. (2002). *Managing across borders: The transnational solution* (2nd ed.). Harvard Business School Press.

Crang, P., Dwyer, C., & Jackson, P. (2004). *Transnational spaces*. Routledge.

Creswell, J. W., & Clark, V. L. P. (2007). *Designing and conducting mixed methods research*. SAGE Publications. https://books.google.co.uk/books?id=FnY0BV-q-hYC

Creswell, J. W., Plano Clark, V. L., Gutmann, M. L., & Hanson, W. E. (2003). *Advanced mixed methods research designs* (Vol. 209). Sage Publications.

Downs, R. M., & Stea, D. (2005). *Image and environment: Cognitive mapping and spatial behavior*. Transaction Publishers.

Faist, T. (2000). *The volume and dynamics of international migration and transnational social spaces*. Oxford University Press.

Glick Schiller, N., Basch, L., & Szanton Blanc, C. (1995). From immigrant to transmigrant: Theorizing transnational migration. *Anthropological Quarterly, 68*(1), 48–63.

Glick-Schiller, N., Basch, L., & Blanc-Szanton, C. (1992). Towards a definition of transnationalism. *Annals of the New York Academy of Sciences, 645*(1).

Goldring, L. (1996). Blurring the border: Transnational community and social transformation in Mexico-US migration. *Research in Community Sociology, 6*, 69–104.

Hall, S. M. (2015). Super-diverse street: A 'trans-ethnography' across migrant localities. *Ethnic and Racial Studies, 38*(1), 22–37.

Jackson, P. (1989). *Maps of Meaning*. Routledge.

Jackson, P., Crang, P., & Dwyer, C. (2004). Introduction: The spaces of transnationality. In P. Jackson, P. Crang, & C. Dwyer (Eds.), *Transnational spaces* (pp. 1–23). Routledge.

King, R. (2012). Geography and migration studies: Retrospect and prospect. *Population, Space and Place, 18*(2), 134–153.

Löw, M. (2001). *Raumsoziologie (Suhrkamp Taschenbuch Wissenschaft)*. Suhrkamp. https://doi.org/citeulike-article-id:355221

Lynch, K. (1960). *The image of the city* (Vol. 11). MIT Press.

Mayring, P. (1994). *Qualitative Inhaltsanalyse*. UVK Univ.-Verl.

Meuser, M., & Nagel, U. (1991). ExpertInneninterviews- vielfach erprobt, wenig bedacht. Ein Beitrag zur qualitativen Methodendiskussion. In A. Bogner, B. Littig, & W. Menz (Eds.), *Das Experteninterview*. Theorie, Methode, Anwendung.

Pries, L. (2001a). *New transnational social spaces: International migration and transnational companies in the early twenty-first century*. Routledge.

Pries, L. (2001b). The disruption of social and geographic space: Mexican-US migration and the emergence of transnational social spaces. *International Sociology, 16*(1), 55–74.

Useem, M. (1995). Reaching corporate executives. In R. Hertz & J. B. Imber (Eds.), *Studying elites using qualitative methods* (pp. 18–39). Sage Publications.

White, R., & Gould, P. (1986). *Mental maps*. Routledge.

Witzel, A., & Reiter, H. (2012). *The problem-centred interview*. Sage.

Yeung, H. W. c. (2005). Rethinking relational economic geography. *Transactions of the Institute of British Geographers, 30*(1), 37–51.

4

Transnationalizing Tokyo: The Gaijin Ghetto

Without going too deep into Japanese history, but giving at least an overview of the first settling of foreigners in Japan and the development of an externally induced foreigners' ghetto in the nineteenth century, this chapter opens the empirical parts of the book by discussing the so-called gaijin ghetto of today which appears to have developed endogenously. The leading question is how this area was selected to become a foreigners' residential area at first, how it has further developed as an exclusive expatriates' "ghetto" and what impact it has on the urban development and society nowadays. Interestingly enough, the socio-spatial concentration of transnational professionals in present-day Tokyo has a long history reaching back to the nineteenth century.

To introduce the reader into an analytical perspective on socio-spatial urban dynamics, particularly on issues of exclusion and other diversification processes, as a thought experiment so to say, the chapter builds its narrative around the notion of ghetto in the classical sense developed by Louis Wirth. In particular, the terminology of the voluntary ghetto, yet, as will be explained below with a societal twist, will be used to explicate the socio-spatial phenomenon of the transnational professionals of today. In this chapter, further urban geographical terminologies that usually

S. Yamamura, *Spatial Diversity in the Global City*, Global Diversities,
https://doi.org/10.1007/978-3-031-64725-3_4

appear in the context of urban segregation, such as ethnic enclaves, gated communities and citadels, will be discussed. This will help to clarify the positioning of the "gaijin ghetto" within the literature and debates on urban socio-spatial segregation. As the first of the three empirical chapters which look into the socio-spatial patterns of the transnational professionals on the local urban level, it follows the previously presented analytical concept of the spacing and synthesis process of producing the transnational space(s).

4.1 From External Exclusion …

Exposure to foreign influences for Tokyo had at latest begun in the mid-nineteenth century. During the era of the Meiji Restoration, when Japan was forced to open its ports for trade after more than two hundred years of isolation (with the exceptions of trading zones with the Netherlands and China),[1] stronger foreign, particularly Western influences flowed into Tokyo. It was not only a mere internationalization in economic terms by which Japan and foreign countries began trading and exchanging goods, but it also encompassed a more comprehensive cultural transformation through the exposure to these foreign influences, resulting in the transformation of politics, society and culture. Foreign Western goods, as well as ideas, started to be integrated into Japanese society and uniquely transformed to fit into Japanese culture; thus, creating a Western style within the Japanese culture which is still referred to nowadays as "yo-fu" in comparison to the traditionally Japanese style of "wa-fu."

The era stretching from Bakumatsu, the end of the Edo period, to the Meiji Restoration was also the time when the first international migrants came to Japan. They were at first restricted to live in the so-called Gaikokujin Kyoryuchi, the Foreigners Settlements. After the arrival of the Perry Expedition with their warships of the US American Navy,

[1] It should be noted that foreign exposures in general were already from ancient times crucial to the development of the Japanese culture and inherent part of Japanese settlement and cultural history. However, the foreign exposure for Tokyo as a specific city can be regarded as occurring only after the establishment of Tokyo as a capital city and its subsequent transformation from a smaller fishermen's town (previously called Edo).

forcing Japan to open its ports to American and further Western trades, these settlements were established in seven treaty ports within the Treaty of Amity and Commerce to allow trades.

In the case of Tokyo, this area was directly at the Tokyo Bay at Tsukiji. The Civilization and Enlightenment Movement (bunmei kaika) for the whole of Japanese society is said to have developed in this area and spread further to the Ginza area. What was first an urban and cultural transformation induced by the personal interest of a daimyo (feudal ruler) to promote Dutch culture developed into a political decision in urban development to form a foreigners' "ghetto" in that area. It is not by coincidence that the first Western-style hotel was established there with the Tsukiji hotel-kan, in which restaurant the origin of the Western-style Japanese cuisine also lies and where the bunmei kaika in general spread from. Nowadays, however, not much of an international flair can be found in these areas; they are, interesting and strangely enough, rather seen as "typically Japanese" areas of socialization by both Japanese and foreign professionals (see further in this chapter). It is an area where the Japanese high society and business people socialize, but which had not been used extensively by foreign peers unless for touristic reasons of experiencing the traditional Japanese Tsukiji fish market or the Ginza shopping streets.

What has become the foreigners' hub in the present day after more than one hundred years since the end of the treaty port concessions and consequently the dissolution of the foreign settlements in 1899 are the areas in the Minato ward, located in the south-west of Tokyo. With a total area of 20.37 km^2 and a population of about one quarter million (Minato City 2023), it is the area with the largest number and density of embassies and consulates within the Japanese capital city. It simultaneously harbors more than 20,000 foreign residents as of 2023. The reason for this geographical concentration of foreign diplomatic missions and accompanying population lies in the urban history following the end of the shogunate.

The dissolution of the Tsukiji Foreign Settlement coincided with the commotion of the Meiji Restoration which followed the end of the Tokugawa shogunate at the Bakumatsu. With the end of the shogunate came the disappropriation of the lands that were occupied by the daimyos.

The feudal lords who were, according to the practice of sankin-kotai, required to alternate between Edo and their respective fief[2] and thus maintain residences in both locations, were disappropriated and the lands were now available to the nascent government (Public Archives of the City of Tokyo 1990). Such lands—especially those of the less powerful regional feudal lords at the periphery of the Yamanote area and the hilly region of the aristocracy (in contrast to the flat marshlands at the Sumida and Tone river delta where the merchants and craftsmen and thus the working class inhabited)—were first allocated to the Imperial military and then with the dissolution of the Tsukiji Foreign Settlement offered to the foreign diplomatic missions for their embassies and consulates. The foreign population shifted gradually from the Tokyo Bay area in Tsukiji at their arrival in the mid-nineteenth century to western parts within Tokyo following the abolition of the foreign settlements in 1899. And this is where the number of foreign residents especially from non-Asian countries is still highest in Tokyo.

Thus, it is by no coincidence that the main concentration of foreign diplomatic missions and the upper-class migrant population is still located in an area to which historically the initially externally (by the Japanese government) excluded foreigners were offered to relocate in the course of the Bakumatsu disappropriation of the daimyo lands. The location of the so-called gaijin ghetto of today, as it will be further introduced in the next part of the chapter, cannot be seen without this urban historical context of a geographical exclusion and relocation of the foreign population. Due to its initially strict restrictions of the residential space given to migrants in the form of the Foreign Settlements but also the "offering" of specific locations during the times of Meiji restoration, the socio-spatial constellation for the foreign migrants can be regarded as comparable with an involuntary ghettoization.

[2] Abundant historical works are available on the function and meaning of this practice, and there are several strategic reasons being discussed. The most important strategic reason behind this obligatory internal circular relocation, however, is the maintenance of the Shogun's power over the distant regional fiefs.

4.2 … To Exclusive Expatriation: The Gaijin Ghetto Phenomenon

> First I lived in, first time I came back as a professional I lived in, ehm, the Meguro area and then I moved into Daikanyama and then I got closer and closer to what I call the Gaijin ghetto. [laughing] Yeah, it's just comfortable. So, you know, when we came back, we lived in Moto-Azabu, Minami-Azabu for eighteen years and it's just, you know, I don't want to live like, I should be honest, I don't want to have the Japanese lifestyle, I want to live a more comfortable lifestyle, near schools. My kids 've been to Nishimachi, so they could walk to school or shopping is convenient, so I don't feel, we need to … we're happy, you know, living in a Gaijin area, it's just comfortable. (DS37)

As the quote above from one of the interviewed transnational professionals illustrates, there is a so-called gaijin ghetto within Tokyo where high-skilled migrants or *expats* socio-spatially concentrate. The term *gaijin ghetto* appeared to be widely used and known among the interviewed professionals and is a self-appointed term (NB: none of the Japanese interviewees nor the triangulated group of interviewees used it). The literal translation of *gaijin* in Japanese means *outside person*—properly translated as *foreigners*—yet it is, as Cybriwsky (2011a: 280) correctly points out, "a casual-language word, perhaps a bit impolite" and has indeed a pejorative connotation. Native Japanese-speakers tend to use *gaikokujin*[3] instead to meet political correctness. Having said this, the further bewildering point in the self-appointed term of *gaijin ghetto* is also the latter part. "Ghetto" is generally associated with the segregation of an economically and socially disadvantaged and underprivileged group: the so-called urban poor, rather than the highly skilled and highly affluent transnational professionals. As it will be clarified further in this chapter, the areas mentioned by these professionals as being part of this *gaijin ghetto* are actually some of the most prestigious areas within Tokyo, with an above-average income of the residents and corresponding high

[3] Literally translated as outside-country-person, which means foreign person in a neutral connotation.

real estate prices. Their ghetto apparently does not comply with the two classical characteristics of the "immigrant enclave," that is, its physical feature to be "less desirable as places to live" as well as the people living there who are recent migrants having "few socioeconomic resources" (Logan et al., 2002: 300).

The question that arises is: why do transnational professionals use a double negatively connoted term for their space? Certainly, there is an ironic, if not sarcastic, undertone to the usage and these words should not be taken literally. And still, it should not be underrated and ignored that exactly these words and none other are factually used to indicate their spaces, reflecting their spatial perception of their Tokyo or at least the part of the expatriates' space within Tokyo. It is, therefore, perfectly legitimate to question its feature by juxtaposing it with conventional ghettos. More precisely, for the purpose of analyzing this particular transnational space, the question on which this chapter evolves is: are there reasonable grounds to call this transnational space a "gaijin ghetto"? In conceptual terms, this means: does the space-making of the gaijin ghetto contain characteristics as attributed to a conventional voluntary ghetto?

Before taking a closer look at this *gaijin ghetto*, a clear definition of ghetto as a phenomenon of social and spatial segregation[4] is essential. Generally speaking, ghetto is understood as "a part of a city in which members of a particular group or race live usually in poor conditions" (Merriam-Webster) or more distinctly: "a geographically definable, frequently discriminated minority quarter after origin, ethnicity, religion or socioeconomic status of homogenous groups" (Lexikon der Geographie, 2001: 49, Vol. 2 [author's translation]), but historically the term dates back to the Middle Ages. There are several etymological theories on *ghetto*, one of the most established being that it derives from the Venetian *gietto* or *il ghetto*, the cannon foundry or respectively the municipal copper foundry, in which vicinity the first Jewish settlement was located (Wirth, 1969: 1–3; Ravid, 2012: 16ff.). Subsequently, Jewish settlements were designated as *ghettos*. However, the usage of the term further expanded to "the area of first settlement, i.e., those sections of the cities

[4] Segregation being a "disproportional distribution of social groups across urban districts or other spatial units" (Friedrichs & Triemer, 2008: 16).

where immigrant finds his home shortly after his arrival in America" (Wirth, 1969: 4) and as early as the 1920s when Wirth wrote his paramount work on "The Ghetto" there was a "tendency to refer to immigrant quarters in general as ghettos" (ibid.). As Ravid points out in his overview of the metamorphosis of the term ghetto, Wirth has become one of the most prominent classicalists to spread the term and extend its meaning in social sciences (Ravid, 2012: 27f.). Although there is some criticism of Wirth having contributed substantially to the vagueness, expansion and over-expansion of the term today (Hoffmann, 2012: 71), Wirth's observation of the ghetto that the urban process involving migrants is basically the "same phenomenon that we see in the social life of other minority groups who live side by side with one another, or, as often is the case, live side by side without one another" (Wirth, 1969: 282) makes the concept of his ghetto still applicable and highly useful for social scientific urban analyses today. The two crucial characteristics to define the ghetto are surely (a) the association with a discriminated and underprivileged population group (cp. Hoffmann, 2012: 53), and (b) its nature of being a socio-spatial phenomenon in cities.

One important aspect has to be anticipated here before continuing with the analysis into the functioning of the ghetto as an organism. Namely, that none of the classical urban models of the Chicago school nor conventional urban theories of the succeeding generations of urban segregations, mostly based on empirical work in the United States or other Western countries, can explain this specific "ghettoization" in Tokyo. First of all, as discussed previously, there is no reasonable ground to compare American and Western cities with Japanese urban development, especially Tokyo with its unique urban history of socio-spatial segregation. On the other hand, the transnational professionals who are the focus of this research are by no means an underprivileged and disadvantaged group of foreign residents. Even if they may have comparable behavioral patterns, such as that they do not invest much in their housing and its maintenance (cp. Fusfeld & Bates, 1984), the reasons are not because of the lack of capital but rather owing to the fact that they, as highly mobile migrants, usually are not focused on permanent residence and, thus, do not invest in permanent housing. Moreover, their housing rents are often not even paid by them but are included in the "*expat packages*" (BJ07)[5] of the transnational corporations for their intra-firm transferees.

Having said so, of course, the *gaijin ghetto* that is lying in front of us is by no means a ghetto in its original sense, especially not as an institutionalized "compulsory ghetto" (Wirth, 1969: 29ff.), but what needs to be further investigated is whether some elements still apply to it. After all, Wirth also writes about the "voluntary ghetto[s]" (ibid. 18ff.) that were created due to "needs and practices rooted in the customs and heritages, religious and secular, of the Jews themselves" already before compulsory acts of authorities resulted in the compulsory ghettos that are being mostly associated with the term. Wirth's designation may be problematic in its contents, but whatever it may be called,[6] there is no denying that there is urban segregation due to compulsion or preference. Such voluntary ghettos were created because "[a] population seeks an area in which it is tolerated and in which the wishes of its members can be gratified with the least amount of interference" (ibid. 286). "Once in the area, each group tends to reproduce the culture to which it was accustomed in its old habitat as nearly as the new conditions will permit" (ibid. 283).

This somewhat provocative approach of analyzing the transnational space of highly skilled professionals in the context of the ghetto concept may indeed also face some skepticism by urban researchers who have been extensively debating and differentiating the similar, yet clearly different, concepts of ghettos, enclaves and inequalities in the city (cp. Varady, 2005). However, on the one hand, the discourse of transnational professionals themselves to designate their living area as a "ghetto"—though apparently paradoxical with regard to their paramount privileged socio-economic background—is already epistemologically well reasoned as socio-spatially subjective, but also collective identification with the spatial constellation is a strong indication of the production of a common

---

[5] There are, in fact, changes in the behavioral pattern including the investment in real estate, see below: Chap. 6 on transitionists.

[6] See Ravid (2012): 33 f. on the discussion of Wirth's terminology. Tych's differentiation of quasi-ghettoization or self-ghettoization opposed to forced ghettoization, as well as the differentiation between a technical or formal ghetto versus non-technical ghetto by Baron, is mentioned in the chapter. Whatever the terminology, the main aspect that there is a difference between these two motifs but they are, in the end, the same had been elaborated by Wirth, too.

transnational space. Also, on the other hand, these well-researched and established related terms do not necessarily apply fully to this specific socio-spatially differentiated pattern and, if so, only at least to the same extent as the idea of the ghetto with its limit of focusing on the voluntary ghetto. By juxtaposing suggested and well-perceived differentiations between the terms of "many forms of clusterings" (Marcuse, 2005: 15), it becomes clear that approaching the term "gaijin ghetto" within the context of Wirth's "voluntary ghetto" is not as absurd as it may appear.

Marcuse, who has explicitly made an attempt to shed light onto the confusion of these urban geographical terms, mainly makes the differentiation between clustering that is involuntary, or better yet, hierarchical (i.e., derived from a ranking system, that reflects superiority based on wealth, status, or power versus those that are "voluntary, that is, nonhierarchical" (Marcuse, 2005: 15).[7] Yet, that this is a different understanding of voluntary and involuntary in the sense of the ghetto terminology is clear. Though it is comprehensible why this distinction is important for him—as he has an urban planning perspective and is concerned with public policy—focuses more on the issue of identifying urban phenomena "of public concern and those of no public policy interest" (ibid.) and thus draws the line there, this idea appears to be untenable, in particular for the transnational professionals' case, because the causal relation is not as easy to differentiate as Marcuse implies. Transnational professionals' residential and socio-spatial choices indeed are voluntary, as they are not confined as in cases of the transnational migrants "from below" with scarce resources or being discriminated against; however, the spatial *clustering* is embedded clearly in a hierarchical order of having such wealth or status to be able to afford housing and specific lifestyles according to the area. As a matter of fact, this particular issue of the voluntary or involuntary nature of the "allocation of space" requires further scrutiny, as we will further see later on, since transnational professionals do not lack monetary resources but may lack resources of

[7] Now, Marcuse's (2005) discussion on which pattern of urban segregation or "residential clustering" is "acceptable" or "objectionable" (p. 5) is another issue which is first to be examined in this part where the characterization and analysis of the urban socio-spatial pattern are the focus, but will be further discussed in the last chapters in the context of the implications that such patterns bear for the urban population and planning.

another kind, such as information or language skills, or be impacted by further external factors influencing their overall decision-making on the choice of residences.

Furthermore, the blunt definitions of the different terminologies, which are potentially related and applicable as alternatives to the *voluntary ghetto* for describing the phenomenon of the socio-spatial pattern of transnational professionals, namely, the enclaves, also do not fully comply with the pattern. The transnational professionals are clearly not parts of ethnic enclaves or an "ethnoracial formation that combines and inscribes (…) 'elementary forms' of racial domination" (Wacquant, 1997: 343) as their spaces are neither limited to the same ethnic group or to a foreign group as still prevalent in the US American debates on urban segregations (cf. Fainstein, 2013; Logan, 2013; Musterd, 2020; Wacquant, 1994), nor is the "means of protecting and enhancing their economic, social, political and/or cultural development" (Marcuse, 2005: 17) necessarily existent. Also, in the same argumentation, the term "exclusionary enclaves" encompasses the aspect of the population group that "excludes others from unauthorized entry" (ibid. 18; cf. Marcuse, 1998). The linear temporary dimension of immigrant enclaves being solely "transitional neighborhoods" for newly arrived migrants due to initial limitations of various capitals (Logan et al., 2002; Massey, 1985) is also questionable in the case of high-skilled transnational professionals (see also Chap. 6 on the phenomenon of transitionists).

Along with concepts of gated communities which are "residential areas with restricted access in which normally public spheres are privatized (…) with designated perimeters, usually walls or fences, and controlled entrances that are intended to prevent penetration by nonresidents" (Blakely & Snyder, 1999: 2), the urban phenomenon of transnational professionals cannot be subsumed under such terms as there is neither an administrative nor a physical exclusion of any population group within the transnational space discussed within Tokyo. Also, it must be mentioned that the strong intrinsic motivations ascribed to the residents of gated communities, such as being "drawn to gates for prestige, image, and status," "want[ing] control—over their homes, their streets, their neighborhoods," or "control in the ability to exclude outsiders from their territory" (ibid. 125) do not apply here and—with regard to the heavily valuing tone—remain questionable, too.

The closest term that may reflect the socio-spatial phenomenon in Tokyo is being suggested by Marcuse, namely, the *citadel*. The citadel is characterized as being "an area of spatial concentration in which members of a particular population group, defined by its position of superior power, wealth, or status in relation to its neighbors, cluster as a means of protecting, displaying, and enhancing that position" (Marcuse, 2005: 26), whereas the means and purposes of such spatial clustering again remain disputable. The focus on "the form of hi-tech, generally hi-rise, mega-projects" (ibid.) is also questionable as the *gaijin ghetto* of the transnational professionals, as we will see further later, is not "separated, insulated, from the city in which it exists" (ibid: 30). Furthermore, though it is also mentioned that these citadels reach beyond the living space of such high-skilled professionals in that there are also "less well-off workers" who assist and serve them and therefore "must be allowed in" (ibid.), what this idea still misses is the multilayered nature of spatiality, that is, that transnational spaces of specific population groups can and do coexist and also overlap with other spaces and that such citadelic places are in fact—neither physically nor socially—exclusive to the population group spatially clustering in them.

Voluntary ghettos or "upper-class ghettos" (Musterd & Ostendorf, 1998) have been criticized as "mischievous; analytically (…) oxymoronic, and policy-wise insidious" (Marcuse, 2005: 16), and the ghetto term, as well as the idea of segregation, has been incited to be used solely to describe "patterns that are undesirable" (ibid.). However, since even the critic's own proposition for "treating other forms of clustering in a more nuanced fashion" (ibid.) appears not to be sufficiently nuanced, excluding aspects as well as imposing means and conscious intentions on the actual individual level of the transnational professionals, and—as a matter of fact—conceptually focuses primarily on residential areas and not the socio-spatial space including leisure and work-related activities, the nuance of *voluntary ghetto* is well justified for the purpose of this work. It is clearly not possible to categorize the transnational space as part of a *clustering* with the race or ethnicity as a "primary factor" as Massey and Denton (1993) propose for the ghetto formation, irrespective of whether or not in an inferior relation to the rest of society. Though Peach (2005) correctly points out the erroneous mixing and confusion of the terms "ghetto" and "enclaves" by the Chicago School,[8] the differentiation still only refers to ethnicity-based clustering and neglects mixed-ethnicity population such as the transnational professionals.

The *gaijin ghetto* is indeed created more under socio-economic or class conditions—normally mentioned in the context of the underclass rather than for the upper class (Wilson, 1993)—and thus it is an example of "structurally imposed phenomena" (Marcuse, 1998: 178), yet there seems to be no clear term for this kind of a voluntary segregation and non-ethnic clustering that helps in analyzing the spatial phenomenon. In fact, urban segregation processes have been extensively researched and described (cf. Friedrichs, 1988; Häussermann & Siebel, 2004) so that the clustering of high-income population itself, in contrast to the urban poor, is not at all a novel phenomenon. However, when it comes to migrants in urban spaces, the analyses have not touched upon the combination of high income and migrant population, and remained silent on the designation of such upper-class segregation in urban space.[9] In the end, as stated previously, the argument that *gaijin ghetto* is a term coined by the transnational professionals themselves contributing to the creation of this specific transnational space as such is crucial enough for further investigation, especially against the background of the underlying spatial concept.

So, in this regard, returning to the actual idea of taking a closer look at this *gaijin ghetto* from the perspective of Wirth's *voluntary ghetto*, it can be

[8] Apart from this critique, though undeniably still a classic for urban sociologists and geographers alike, the Chicago School has been heavily contested on further issues, too, including the main criticisms of its neglect of choices and preferences in the socio-spatial patterns (Hollingshead, 1947) as well as institutional and political factors (Denton & Massey, 1991; Van Kempen & Murie, 2009; Van Kempen & Şule Özüekren, 1998), but also on the descriptive nature of its models with limitations for empirical applicability, especially beyond Northern American and capitalist cities (Lichtenberger, 1986), or its neglect of vertical dimension for differentiation (Carter, 1977). However, for the purpose of this chapter, which is mentioning and using the ghetto of Wirth solely as an analytical framework for approaching the discussion of the existent empirical data, the critique and legacy of the Chicago School will not be further delved into.

[9] Though dealing with quite different urban populations, as the extent of segregation in German cities is rather incomparable to that in American ghettos, Häußermann and Siebel (2004: 159) also mention the differentiation between voluntary and involuntary segregation.

questioned whether there are elements of the transnational professionals' production of transnational space similar to the reasons for which voluntary ghettos were created. Are there reasons why transnational professionals find this area "just comfortable" and allowing them their own "*more comfortable lifestyle*" (DS37) in contrast to the rest of the too-Japanese Tokyo, as quoted previously? Is there, for example, "the element of a common language, of community of ideas and interests, and the bare congeniality that arises even between strangers, who, coming from the same locality, meet in a strange environment" (Wirth, 1969: 20) that results in the production of the gaijin ghetto? If "every people and every cultural group may be said to create and maintain its own ghetto" (Park, foreword in Wirth, 1969: vii), then transnational professionals, especially from the same industry with comparable career backgrounds, would be likely to create their own ghetto, too. Even though we may not find the general features of the ghetto as a segregated area of the urban poor, will we be able to still find a ghetto in terms of "a cultural community that expresses a common heritage, a store of common traditions and sentiments" (ibid. 289)?

According to the underlying spatial concept of this research (see Chaps. 2 and 3), we should bear in mind that we are not only looking at the ghetto as a cultural community but also their direct and intrinsically tied production of the ghetto space in its geographical extents. The close connection between the urban phenomenon as such and the socio-cultural character of the people producing it is actually the crucial element of the underlying concept of transnational space.

As we will see, the approach to closer investigate the *gaijin ghetto* from the perspective of the transnational space as an analytical means results in remarkable insights into the *constitution* and the *synthesis process* (Löw, 2001) by this particular group of transnational migrants. Furthermore, the analytical approach to initially view the transnational space, which is conceptually embedded in a macro-perspective debate of global transnationalization of economy and society, from the urban-level ghetto lens also reflects what Wacquant suggested as an analytical framework and necessity, which means an "institutionalist (i.e. relational) conception of ghetto" which "foregrounds and interrogates variables that more nominalistic or gradational approaches tend to treat as background conditions"[10] and empirical research has been neglected (Wacquant, 1997: 344).

After all, as Wirth says about his interest in Western European and American ghettos, the "gaijin ghetto" of transnational professionals in Tokyo may, just as Wirth's ghetto,

> [illustrate] picturesquely the ways in which a cultural group gives expression to its ancient heritage when transplanted to a foreign setting, the constant sifting and resifting of its members, and the forces through which the community maintains its integrity and continuity. (…) [T]he ghetto demonstrates the subtle ways in which this cultural community is transformed by degrees until it blends with the larger community about it, meanwhile reappearing in various altered guises of its old and unmistakable atmosphere. (Wirth, 1969: 5)

To analyze the constitution of the gaijin ghetto space according to the analytical framework based on Löw and presented before (Chap. 3), the following section starts with geographically locating the area described by the interviewees as being the "gaijin ghetto" within the vast Tokyo metropolitan area. Then, as a second and main step, drawing from the information in the accounts of the transnational professionals themselves, descriptions of the areas will be presented. The corresponding analysis encompasses several aspects: the description of spacing elements, that is, characteristics of the space with specific institutional, symbolic and other space-structuring elements in this area, followed by the social practices taking place in these areas which contribute to the actual socio-spatial space-making, and finally also the synthesis process of the transnational space, that is, associated and interpreted characteristics of the area itself. This will then bring us back to the question of the context in which the transnational space of these migrants is embedded and whether their socio-spatial patterns can be actually paralleled to a voluntary ghetto as supposed or may also have aspects that constrain the transmigrants in their socio-spatial choices.

[10] Wacquant, as mentioned before, primarily refers only to the ethnic and racial segregation and to those who have been under societal domination, and does not consider voluntary and upper-class clustering in his treatise.

4.3 Localizing the Gaijin Ghetto Within Tokyo

Speaking about the *gaijin ghetto*, the first aspect is to locate it geographically within Tokyo. As it can be seen in Fig. 4.1, the high-income population, which means income above the average of the twenty-three metropolitan wards (i.e. 4,621,000 JPY[11]), is concentrated in six central to western wards. The high-income household are concentrated in Minato, Chiyoda, Shibuya, Chuo, Setagaya and Meguro, whereas Minato with an average annual household income of over 11,846,000 Japanese Yen leads the group. The top income wards are followed by Chiyoda (9,851,800 JPY) and Shibuya (9,111,6500 JPY) (Official Statistics of Japan, 2021).

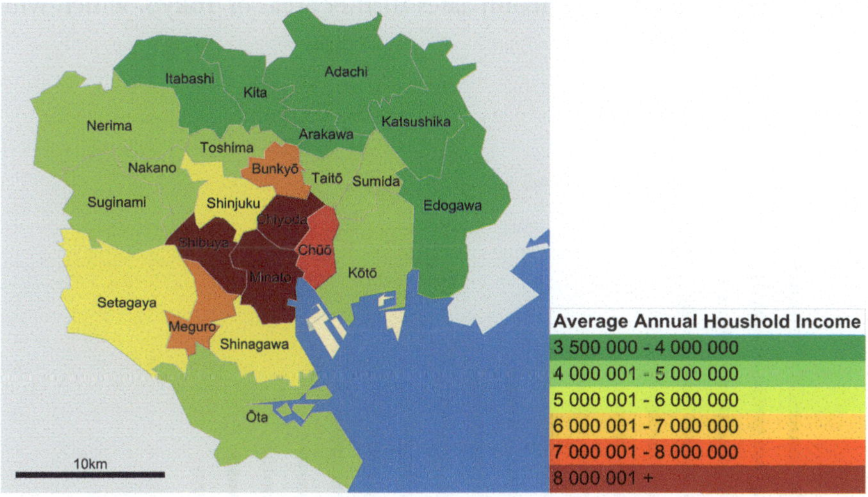

Fig. 4.1 Overview of household income in Tokyo 2021 (author's own, based on data provided by e-Stat Official Statistics of Japan)

[11] The national average income in Japan is at about 2.7 Mio JPY, contrasting the already above-average income level of Tokyo. However, as the statistics clarify, the socio-economic polarization within Tokyo is high, too. While these numbers may fluctuate strongly due to current inflation rates, the point that is clear here is the dimension of socio-economic privilege of the particular group of transnational professionals we are focusing on.

At the same time, with no direct statistical correlation so far, the general foreign population within Tokyo is concentrated in other wards (Fig. 4.2)—the top five being Shinjuku, Edogawa, Adachi, Koto and Toshima—with only Minato and Setagaya following within the top ten. So, purely statistically, there is apparently a divergence of the foreign population on the one hand and the high-income population on the other, whereas the peer group of this study is covering both criteria. If one thus looks at the proportion of the foreign population in the high-income wards in the total foreign population of Tokyo (Fig. 4.3), it becomes quite clear that the transnational professionals that we focus on in this work, contributes to a small yet economically highly affluent minority of migrants.

As the interviewee in the quotation at the head of this chapter already mentioned, the Azabu area—among it Minami-Azabu and Moto-Azabu, both parts of the Minato ward—plays an important role in the "gaijin ghetto." However, it also encompasses further areas of Tokyo, which will

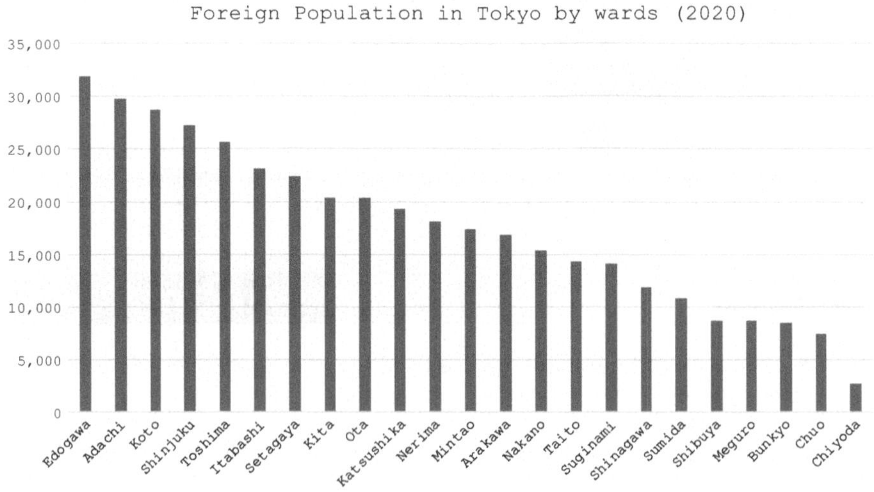

Fig. 4.2 Foreign Population in Tokyo (23 wards) in 2020 (author's own, based on statistics by e-Stat Official Statistics of Japan)

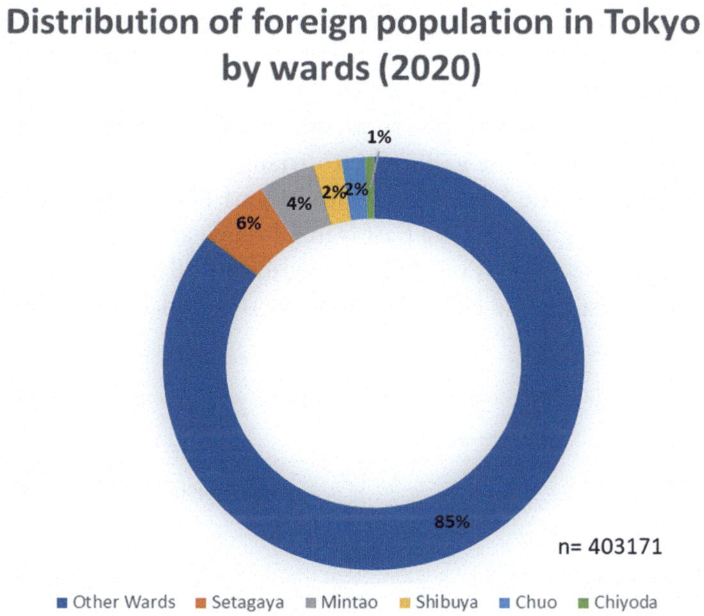

Fig. 4.3 Distribution of foreign population in 2020 (author's own, based on statistics by e-Stat Official Statistics of Japan)

be presented here.[12] Functionally, the "gaijin ghetto" was mentioned as an area of living, that is, for residential and socialization purposes such as meeting other transnational professionals, running daily errands, and eating or going out with family and friends as well as semi-formal after-work socialization with co-workers. It must be noted that the gaijin ghetto is an integral part of the overall transnational space but only a special area within, with a strong focus on the function of residing and socialization; functions such as work are not necessarily covered by the gaijin ghetto that the interviewees referred to.

Turning now to the actual geographical extent, what has been mentioned a few times in the interviews with the transnational professionals and also confirmed by one foreign real estate agent specializing in

[12] As a matter of fact, if one takes a look at the distribution of the foreign population in Tokyo, it becomes clear that the districts to which transnational professionals refer to are areas with only a few foreigners as a share of the total foreign population within Tokyo.

expatriate housings is the "triangle" form of the gaijin ghetto (Fig. 4.4). The size of the gaijin ghetto, respectively the spatial expansion of the triangle, slightly differs depending on the eye of the beholder, but there are elements recurring, such as specific areas and subway lines (Fig. 4.5). The narrower definition of the gaijin ghetto triangle stretches merely over an area of 3.7 km^2 at the borders of Shibuya and Minato wards. The outposts of the gaijin ghetto to the East is in Chiyoda at the Central Business District and to the affluent residential neighborhood around Denenchofu to the south-west, in the Meguro ward—as the reader might remember,

Fig. 4.4 Gaijin Ghetto within Tokyo (with Golden Triangle) (author's own)

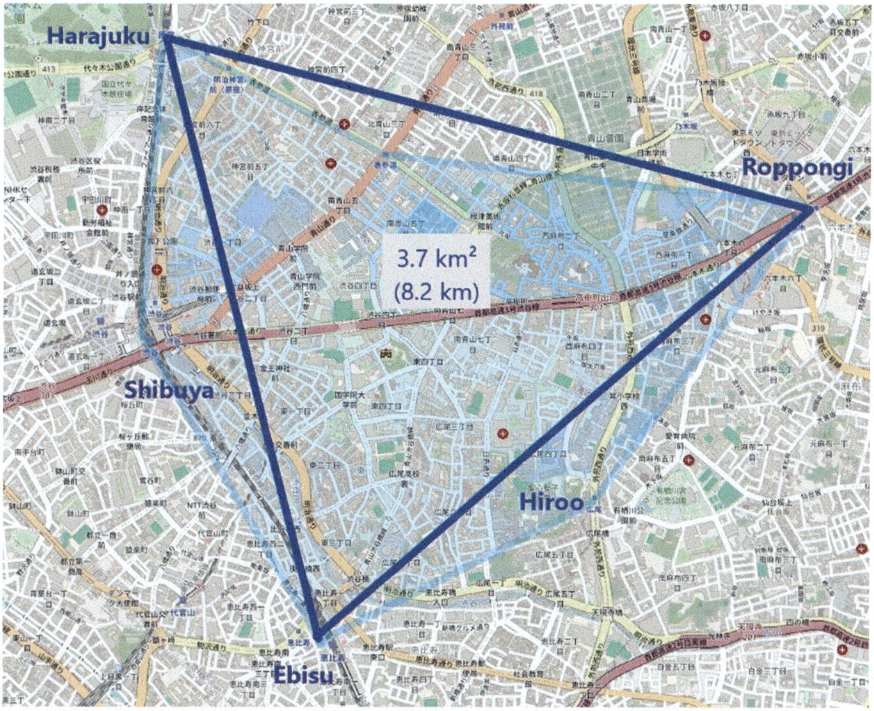

Fig. 4.5 Extent of Gaijin Ghetto (author's own, based on OpenStreetMap)

all in the top group of wards with the highest annual income households in Tokyo.

When looking at the concrete places that have been referred to as essential parts of the gaijin ghetto, the area that has virtually always been mentioned is Roppongi, the farthest eastern apex of the gaijin ghetto triangle, sometimes stretching to the adjacent eastern area of Kamiyacho and Atago. The Roppongi area, which has also often been mentioned in literature (cp. Cybriwsky, 2011b), appears to be the epitome of the gaijin lifestyle and its ghetto:

> I always end up being in Roppongi for some reason… Because there are all the 'gaijin' people hanging around! Yeah, the expats. I mean, people seem to think Roppongi is the center of Tokyo. (RM01)

And:

> (…) there is Roppongi, right, it's famous for all the bars and clubs and everything like that. So, most foreigners like to go there partying at first… (…) So in Tokyo there is definitely a sense of that there is, you know, some bars in Roppongi that you go there, you'll see a lot of familiar faces every time you go, that you … you hear people talk about oh, was in such and such bar. (RS03)

The other vertex of the gaijin ghetto triangle is the Azabu area and/or the Hiro'o area. Sometimes, Hiro'o and Azabu are regarded as the two vertices, sometimes perceived as the same area[13]:

> That was, that was a conscious choice on my part, because when I arrived there back in '88 everyone was living in what was called the gaijin ghetto, which was over near Hiro'o. (BM22)

The third vertex is rather diffuse; the triangle leg can stretch from Azabu/Hiro'o further to the north to just below Aoyama or the Shibuya/Omotesando area. What has been often mentioned as being an area strongly frequented by foreign expatriates is Ebisu, farther in the western part from Azabu along the Hibiya line. This, however, is usually not regarded as within the triangle but rather an extension of one of the legs.

In the context of the spatial expansion of the gaijin ghetto, attention must be drawn to the fact that public transportation, in particular the subway, plays a prominent role in the daily lives of high-skilled migrants

[13] There is some confusion and misconception of the area "Hiro'o." It is often designated as the area with "many consulates" (both online on expatriates' sites as well as in literature, see Cybriwsky (2011a)); however, correctly, there are only five consulates in Hiro'o. Most of the consulates are actually located in the adjacent Minami-Azabu which is a part of the Minato ward across the Meiji-dori, in contrast to Hiro'o which is administratively a part of the Shibuya-ward and covers only the west side from the Hiro'o subway station. When people refer to Hiro'o colloquially, the area covers both sides of Hiro'o and stretches over both wards. According to a local Hiro'o resident, the south-west part used to be called Moto-Hiro'o, compared to the newly developed area on the east side of the Meiji-dori, that is, now Minami-Azabu, called Shin-Hiro'o (new-Hiro'o). Adhering to statements by local Hiro'o residents as well as ward officials well acquainted with the locality, it can be put to record that the area within approximately 500 m of the Hiro'o station is commonly perceived and colloquially named as Hiro'o. Thus, when people refer to Azabu area, it often includes the Minami-Azabu area as well, which in turn often covers the Hiro'o area and vice versa.

in Tokyo (Fig. 4.6). In contrast to some other global cities where public transportation is a means for the working class rather than the upper class, the use of public transportation is common in all socio-economic strata. In particular, the subways, namely, the Tokyo Metro and Toei Subways, are important means of transportation. The subway lines and the stations are integral parts of the daily lives to the extent that they are almost always reflected in the mental maps of the transmigrants. The subway stations especially have emerged as important pillars in the spatial order. When places are mentioned, the migrants referred most of the time to the area around the subway station with the place name rather than the administratively territorialized areas. As a matter of fact, attention must be drawn to the divergence between administrative boundaries of the metropolitan area and the common usage of place names. Tokyo as such is still a fairly young administrative entity and even more so are the 23 ward boundaries, which were introduced in the post–World War II era. Areas were defragmented administratively and since people not only internationally but also within a city move beyond borders and create social spaces crossing administrative border, the spatial orientation focuses

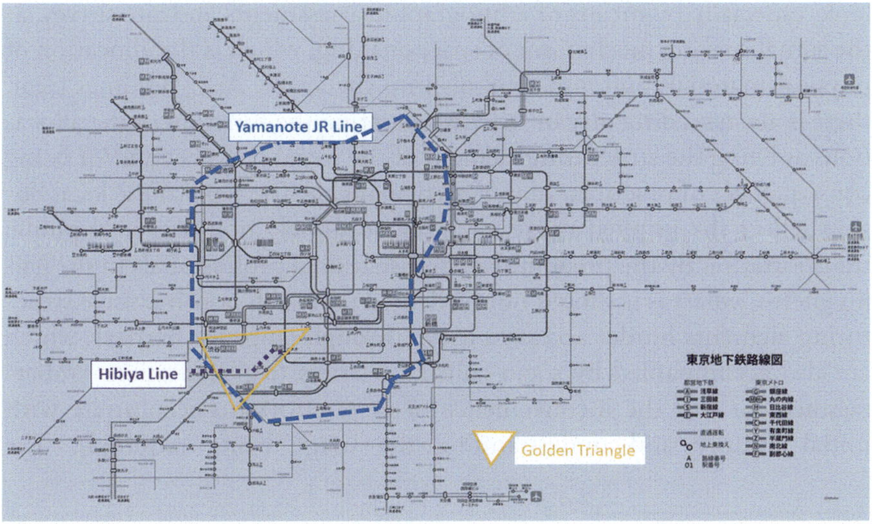

Fig. 4.6 Tokyo Subway map with references to the gaijin ghetto (author's own, based on Wahsaw, 2019/Wikimedia Commons/CC-BY-SA-4.0)

on subway stations and their neighborhoods which constitute the socio-spatial behavioral patterns.[14]

Concretizing the gaijin ghetto area by means of the transport system, the circular Japan Railways (JR) Yamanote-line emerges clearly as a significant pillar for geographical orientation/localization in the mental maps of the gaijin transnational professionals. The interviewees showed strong tendency to focus their spaces of activities along the Western stretch of the Hibiya line, namely, the stations Kamiyacho, Roppongi, Hiro'o and Ebisu and occasionally Nakameguro (see subway map above: *green dotted line* = Yamanote line, *yellow broad line* = Hibiya line). Though the subway as public transportation is a structuring element of the transnational professional's socio-spatial patterns, it is interesting to also note that other means, such as the local bus system, appears to be non-existent in the discourse. This peculiar observation is also underlined by the local bus station where information in an otherwise multilingual, particularly English-dominated, area is limited to Japanese with scarce and more generic translations of the departure and destination stations of the bus lines in English, Chinese and Korean (Fig. 4.7).

The latter aspect also emphasizes the significance of taking a closer look, especially by means of ethnographic assessments and site survey, at the actual spacing mechanism of space-making, which is the allocation of physical materials and also symbols into the space. As it becomes clear, there is always a difficulty in strictly separating the function of the allocations as being the structuring element of the social practices and/or being consequences of practices. In this particular case, the signs at the local bus stop reflect the implied usage, or more accurately non-usage, of public transportation by the predominant transnational professionals in the gaijin ghetto, whereas it should not be denied that such non-inclusive structuring elements hinder access to usage through language barriers, which are also accompanied by socio-cultural barriers (e.g. hesitation, embarrassment or just the inconvenience of extra efforts to acculturate with initial inquiry and understanding the usage being required). The

---

[14] Evidence for this tendency to station-centrality in the socio-geographical orientation of Tokyoites are surveys of most livable areas where stations are used, designating a specific area's flair and environment, rather than using area names.

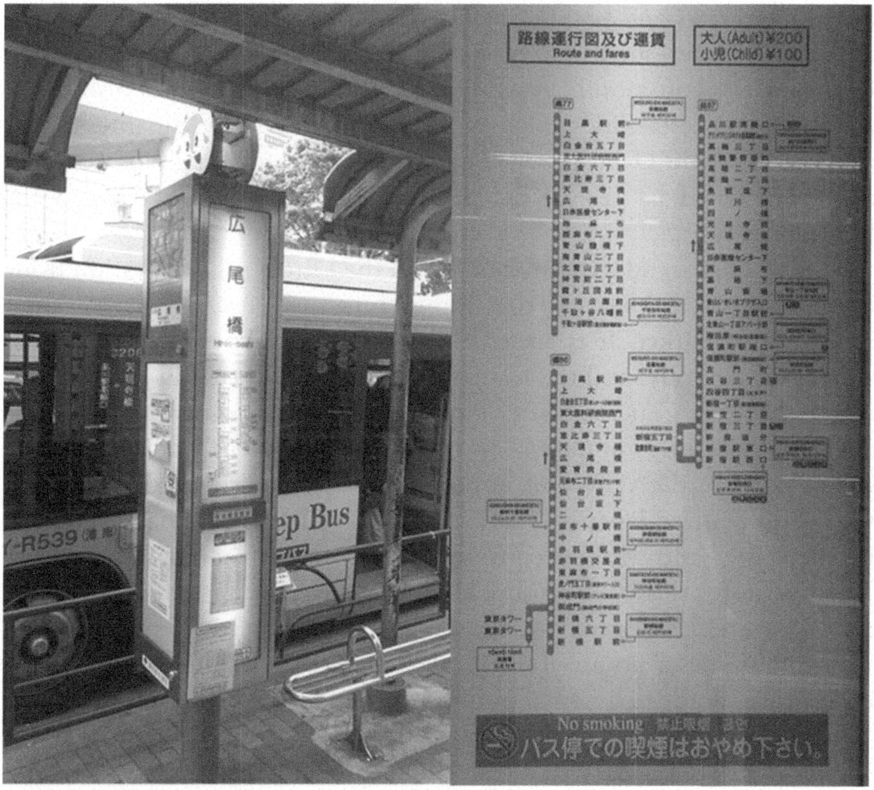

Fig. 4.7 Bus stop Hiro'o-bashi (author's own)

following part of the chapter, thus, deals with the allocation of physical materials into space by providing the descriptions given by the interviewees of their gaijin ghetto.

4.4 Spacing: What Can Be Found in the Gaijin Ghetto?

Important sites for daily life in this gaijin ghetto and simultaneously a crucial reason for the spatial concentration are the housing, which allow a Western-style living, usually without a Japanese tatami mats room, kitchen interiors and bathing facilities according to Western standards, and also features such as a representative dining room for receptions which are less common in typically Japanese apartments. But also shopping and service facilities, such as the National Azabu (Minami-Azabu) and Nissin World Delicatessen (Higashi-Azabu), are of high importance. Also mentioned by several people, but farther away, was Costco (in Kawasaki-city).[15]

One of the first advertisements within the Hiro'o subway station from the conducted site surveys that jumps into one's eyes is the National Azabu Supermarket. It is aimed clearly at English-speaking customers, with opening hours and the route to the store all written in English and even pictures of Caucasian and other "non-local" customers from diverse ethnic backgrounds emphasizing the customer base. Characteristics of such stores are the variety of imported goods, English-speaking staff as well as Romanized Japanese[16] or even English labels on virtually all goods in the store, and with accordingly higher prices compared to local Japanese stores (Fig. 4.8). It should be noted that despite the wider variety of imported "Western" goods, this supermarket also offers local foods with English descriptions of the product (Fig. 4.9). In line with the high-end service, home delivery of purchased goods is also being offered.

Not mentioned by the interviewees themselves, but apparent elements of the "comfortable lifestyle" as in the introductory quote that could be observed at a complementary site survey in this area were services such as medical and pharmaceutical facilities (Figs. 4.10 and 4.11), and also mobile telecommunication services and restaurants providing English and multilingual assistance, support explicitly aimed at international, particularly English-speaking, clients (see Fig. 4.12).

These services and businesses being offered and located in the gaijin ghetto contribute to the comfortable lifestyle for transnational

---

[15] Interestingly, there are more high-end stores in Tokyo, such as Kinokuniya International Store in Aoyama and Meidi-ya in Hiro'o, with imported goods but they were not mentioned by the interviewees. Also, observations in the Meidi-ya showed that there was no apparent English-speaking customer orientation as in the National Azabu despite the variety of imported goods, with Romaji/English descriptions solely on the meat counter as well as ham/cheese within the store.

[16] Romanized Japanese: Japanese words transcribed into the Latin alphabet

Fig. 4.8 Linguistic landscape in gaijin ghetto supermarket (1) (author's own)

professionals. As one of the interviewees explains the core characteristics of the gaijin ghetto or "expat bubble":

> Interviewee (KW45): Yeah, yeah, we definitely didn't stay in the bubble. Akasaka....
>
> Interviewer: What do you mean by the bubble?
>
> Interviewee (KW45) Oh, people call it the expat bubble. If you want to, you never have to interact with anybody Japanese. You can go to the American schools, there's two American...I call them American... there's two Western grocery stores. Ehm, you don't ever have to explore, but we were really quite good at it.

Fig. 4.9 Linguistic landscape in gaijin ghetto supermarket (2) (author's own)

In a brief interview, the staff in one of the mobile service stores explained that these English-language advertisements and offers are not large corporate strategies directed from the national headquarters but rather adaptations to the very local market, where strategies were decided and implemented locally by the staff (independent from other stores in the same area, but solely individual decisions of the respective shop). At the same time, contracts in English were provided by the headquarters to the retail stores, as well as marketing stickers for services provided nationally (see above photograph regarding "interpreting service through TV phone now available").

One prime institution for the socialization of these professionals, with and without family, is the Tokyo American Club (TAC), mentioned by

Fig. 4.10 Pediatric clinic in Hiro'o (author' own)

the majority of interviewees, either as active users or as observers of other's gaijin-lifestyles. The TAC was originally founded in 1928 and in its "sixth incarnation" provides sports facilities, restaurants and facilities for other leisure activities exclusively for members, as well as an area for invited guests of members for business meetings. To illustrate the exclusivity of the club, which has the US ambassador to Japan as the honorary president, a brief self-description of its offer on the homepage: "The building was described by private club consultants the McMahon Group as 'light years ahead of its U.S. counterparts' and "[as] quite possibly the finest private club facility in the world. Members enjoy access to world-class recreation facilities, including a roof-top pool, bowling alley, golf simulators, full-size gym, library, childcare center and spa, as well as a

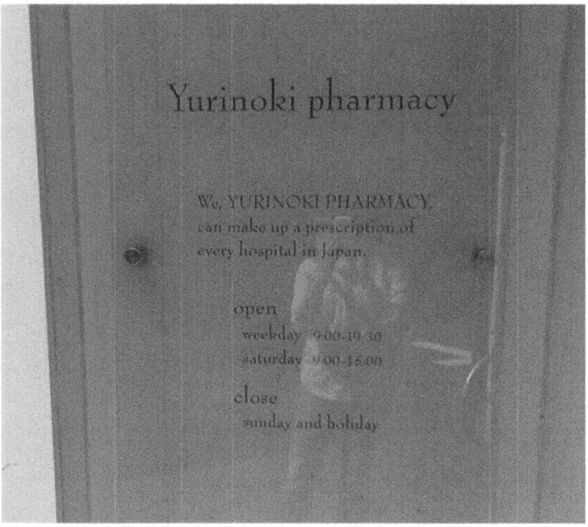

Fig. 4.11 Pharmacy in Hiro'o (author's own)

host of fitness, cultural and educational programs for all ages. Besides being home to five restaurants, a bar and a seasonal cafè, the eight-story facility features seven overnight Guest Studios and superlative meeting, party and conference facilities." (TAC).

Other significant institutions in these areas are educational and socialization facilities. The locations of international schools—and also stops for the school busses to international schools—which are partly in the suburbs, for example, the American School in Japan (ASIJ), are a crucial aspect for transnational professionals with children when choosing their residence. At the same time, the real estate agencies—being aware of these preferences regarding the schools—focus on the bus stops when marketing their housing for professionals by providing maps of the inner city with international-customer-based shops, schools and their bus lines. This once again illustrates the complex interrelation between the structuring spacing aspect and the according structured social practices on the other hand as both are results and causes for the socio-spatial patterns. In this context, it should once again be reminded that the density of consulates and other public institutions in this area is presupposed by the historical background in the Meiji era.

Fig. 4.12 Telephone communication service shop in Hiro'o (author's own)

Summing up this context quite adequately with the causality dilemma of the "chicken or the egg," one of gaijin ghetto professionals said:

> Well, I think, you know, first of all there's, I don't know which comes first, the chicken or the egg, but if you want to live in a luxury house or apartment, you know, one that's large and has, you know, all the kind of amenities that you, you know, that you're used to, ehm, you know, the greatest concentration of them are in the Azabu area, you know, some are now, yeah, and there's another Gaijin area, kind of [unknown]. But, so if you will prefer, you know, a certain type of apartment, you know, you're looking

> for apartment that is at least, you know, whatever 200 square meters and has top appliances and several bedrooms and so forth, you know, the greatest concentration of them is in the areas and then, the schools are nearby, the shopping is nearby, the American club is nearby, I work in, ehm, nearby, in, near Miyako Hotel, Ark Hills. So, ehm, yeah, everything is, there's the greatest concentration of housing and the other amenities and schools and that's what attracts people to these areas. (DS37)

4.5 Constitution of Gaijin Ghetto: What Makes the Gaijin Ghetto?

The second aspect to be discussed now is the social practices which are connected and embedded in these spacing processes, by which the gaijin ghetto becomes not only an agglomeration of specific businesses and services and accompanying signs in public space but also constituted as socio-space. It is by social practices of transnational professionals in these areas that the gaijin ghetto obtains its meaning as a socio-spatial phenomenon of the ghetto itself. Social practices encompass not only behavioral habits such as shopping or residing which are carried out by individuals—though in a collective pattern—but also interactive practices, meaning social interactions such as dining out, meeting for leisure activities or "partying" as mentioned previously regarding Roppongi.

Regarding the social environment in Roppongi, the following quote clarifies that a cluster of specific professionals can be found:

> And it's predominantly people that work in the financial industry or at least in an international kind of a, a company. Rather they work for finance, financial sector. They also meet there. So I would say within those places, you tend to meet the same people, usually, and, and there's very few domestic people. Like, sure, there are domestic individuals, but predominantly there are, like, the international individuals—you know, at least people speak English. Even the bartenders speak English, so it's very … it's geared towards people that are, you know, from…you know, the expats, mainly, I think. (TK11)

As it can be assumed, the language issue is dominant in the accounts of the transnational professionals. Here, several made the distinction between global cities that they had been dispatched to and made a clear differentiation of Tokyo as a global city:

> I, I think I was already pretty global. Which I guess in, in Japan was also different, because there, English is not the, the common language, which meant that all, all foreigners, all non-Japanese, almost formed a group together, because they had the same problems with the language, the same problems, they couldn't read anything in the, the supermarket etc. Whereby here, since Singapore is English-speaking with people from Great Britain, from US etc. For them, they don't feel so separated from the rest of society, just because the language barrier is already not there. Historically, Singapore is anyway, was a British colony and has a lot of English influence though. So I think here, the, the non-English-speaking nations form their own groups, whereby in Japan I, I had the impression it was more like all non-Japanese formed their own group. (IN04)

Again, the language is an important aspect and this distinguishes the Tokyo case from most other global cities, which should be kept in mind for later discussion. Tokyo remains a unique environment compared to other Western and also Asian global cities as Singapore or Hong Kong, where English is spoken; yet, non-English-speaking global cities are and will be emerging in future. It should, however, be also noted that the interaction with English-speaking people is the actual key to this socio-spatial pattern and not necessarily the sole language environment itself. After all, there were many professionals who had acquired Japanese skills over time and even married a Japanese spouse but remained in the gaijin ghetto. As a matter of fact, even Japanese spouses can be detached from the local people despite language proficiency, as described in the following case:

> One interesting thing is, when…my son's in junior high school now, when he was in elementary school, the soccer team was very much for the parents. All of the elementary school kids, the soccer mums all attended the games and some of the dads, but all of the mums certainly, cheering and going crazy and this was a big event for the mums. (…) So, my wife doesn't

fit in with the soccer mums, it's not good with other Japanese parents, because, I don't know why it is, but as Japanese as my wife can be on some things, she's changed a lot from living overseas. She…" (I: [81:37] "Ok, did she grow up overseas?") R: [81:39] "No, no. Not at all, but from the time, you know, we were married and my kids were born, all the way up to 2006, we lived in various places, you know, in America. And, ehm, I don't know, I think my wife doesn't fit in very well with the other soccer mums, they don't have the same, obviously their husbands are from different places and their experiences are different, they just don't bond very well. (NB25)

Similar accounts of not fitting in to the local society, and by that being drawn more into the gaijin ghetto-type places, can also be observed by partly Japanese professionals (mixed ethnic background) who are proficient in the language but explain their preference for English-speaking persons:

My friends tend to speak English quite well, because they had, they were abroad, they've been to international schools or there seems to be a large part of my friends are in this sort of English-speaking community, yes. I don't find many, I don't relate as well to groups of pure Japanese people because although I find it really fun, I don't understand the nori[17] sometimes. (KL41)

The actual importance of the direct social environment over the language issue itself is also illustrated by the professionals' accounts of where they socially interact and with whom in an after-work semi-formal to informal setting. On the one hand, the location of facilities, such as the company itself, plays an important role:

After-work drinks, you know, we might go to, my office was in Roppongi 1-chome, so we go to some local place there, you know, just for a drink. If we have a team night out, we might go to Izakaya somewhere or I don't know. It could be anywhere. We wouldn't go too far, so basically because

[17] Nori (code-switching into Japanese): Japanese word for humor or situational mood

> the office were in Roppongi, we would always either be in Roppongi or Akasaka. (JB23)

But also locational facilities, such as the accommodation of the business partners, play a significant role:

> So and if we go to a restaurant and the guy … sometimes some of them do not know their way around Tokyo so the easiest one is do something close to the hotel so the guy can go because it is their hotel and they can get home. So, chicken and eggs whether you are going to Roppongi, Akasaka, Nishi-Azabu, Aoyama. It is still very much this sort of gaijin ghetto, I guess. The guys who lived in Tokyo a lot longer and still living here we go much, much further. (JP43)

What becomes clear is that socio-spatial patterns are adapted according to the social contact. In the previous case, a person who also ventures "much further" from the gaijin ghetto stays in the ghetto to accommodate the needs of, or provide convenience to, the social contacts that are—in his view—less familiar with the city.

Parallel to the momentary adaptation of one's socio-spatial pattern according to the social contact through the work environment is also the social environment of the transnational professional himself/herself. As mentioned already, a convenient connection to international schools is one of the main selection criteria when searching for housing. Such choice of socio-spatial pattern contributes to the production and also reproduction of transnational spaces in the form of the gaijin ghetto. It is not only the transnational professionals who interact according to this pattern, but also their children, which further is amplified by the social environment the children are embedded. By being involved not only in the transnational corporate business networks, but also in a specific social network of international schools or leisure activities with similar, if not the same type of transnational professionals, the socio-spatial pattern begins to densely overlap, and contribute to an even dense transnational space of the gaijin ghetto.

As it can be already assumed, socio-spatial patterns can drastically be altered due to changes in the life course situations. Compared to a single young professional on an overseas assignment, a married couple certainly has a different social pattern as would a transnational professional with children do. Such dynamic changes in the socio-spatial patterns will be discussed later in the chapter on transitionists.

4.6 Synthesis Processes: How Is the Gaijin Ghetto Interpreted?

The third aspect of the transnational space-making that needs to be clarified now is the synthesis process of the gaijin ghetto. The gaijin ghetto is quite precisely located within Tokyo and the characteristics of the location itself with the businesses and also the social practices within have been presented. What remains is how the gaijin ghetto is being processed in the minds of the transnational professionals. As explained in the methods chapter (Chap. 3), mental maps are one means to make such synthesis processes visible.

Interestingly, the mental maps drawn by transnational professionals with the gaijin ghetto patterns actually reflect the synthesis process quite clearly. They imply that the spatial localization of the "gaijin ghetto" explained previously is also memorized, imagined and perceived as such in their interpretation of the transnational space. It must be noted that mental maps by their nature do not necessarily, and more seldomly, correspond to the actual geographical maps, as these only reflect the image of the urban area of the respondent; also, due to the linguistic barrier, the names of the locations drawn onto the mental maps are, in the case of Tokyo, often not correct. However, what the mental maps are able to show is the idea of the spatial surrounding of the respondents' action space within Tokyo. The clear pattern in the mental maps is—with regard to the gaijin ghetto—the rough triangular shape or trinity of the space, such as follows (Fig. 4.13):

This triangular shape is actually not only mapped out by drawing but also clearly memorized as such in mental maps which were vocalized and sometimes even distinguished as follows (bigger triangle dashed in orange,

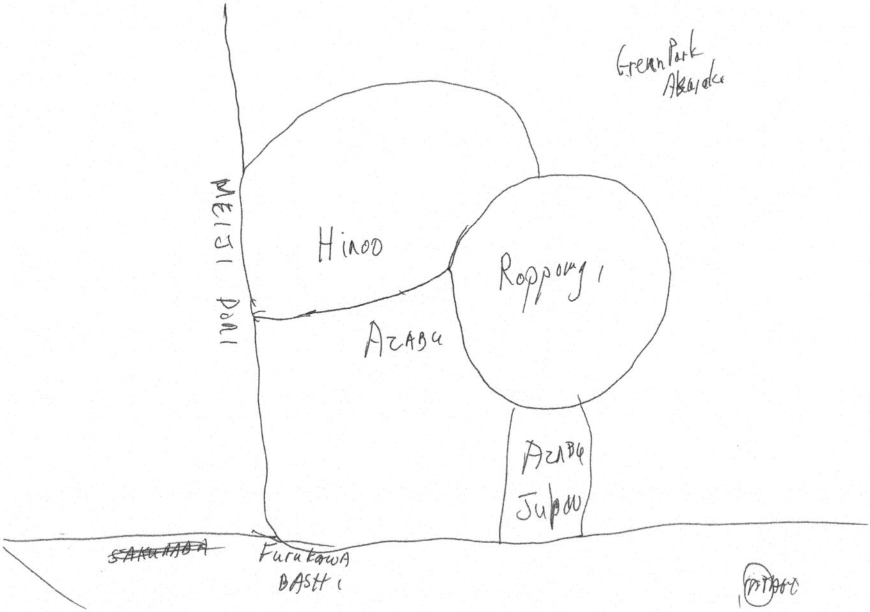

Fig. 4.13 Mental map of gaijin ghetto (research material of author)

smaller triangle continuous line in blue [as added by author for emphasis], see Fig. 4.14):

> Interviewee (JP45): I am probably still in my zone. But I mean there [pointing at what is drawn to be the location of Roppongi] and kind of Aoyama to Ebisu-Hiro'o, they are probably like in a smaller triangle. I am in a slightly bigger triangle but it is inside that.
>
> Interviewer: Okay, at the bigger, right? [pointing at the larger triangle on the map]
>
> Interviewee (JP45): Yeah. That is the golden triangle. [Chuckles]
>
> Interviewer: Golden triangle being Roppongi, Hiro'o and …
>
> Interviewee (JP45): Almost everywhere. Harajuku down to Ebisu and then from Harajuku kind of to back here [work around Marunouchi]. It is probably like that.

Furthermore, the gaijin ghetto as a transnational space, though geographically not necessarily consistent with the vicinity to work and not necessarily coherent with the other local social environments within

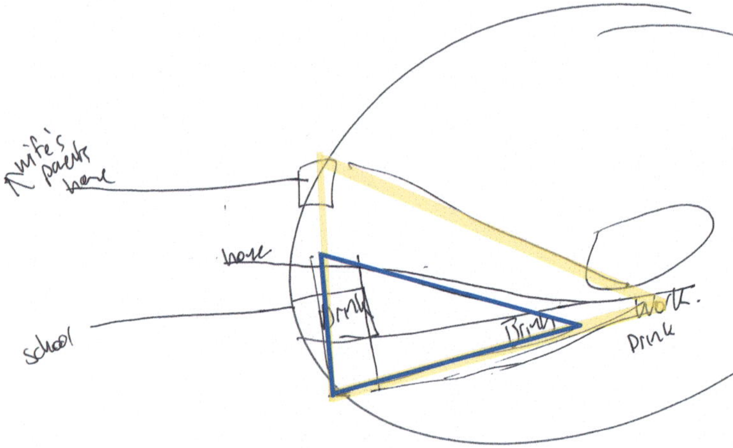

Fig. 4.14 Mental map of the golden triangle of gaijin ghetto (research material of author)

Tokyo, is strongly associated with the idea of a comfort zone mentioned at the beginning of this chapter. It is primarily the language, the social environment and simply the "comfort" that is being associated with the Golden Triangle:

> So it is convenience. For foreigners who do not speak Japanese, this is the one area where you might be able to have some kind of a lifestyle without using Japanese. That prism here is a little bit easier. You have got your National Azabu—and your comfort zone. (JP43)

4.7 The (In)voluntary Nature of Transnational Professionals' Ghettoization

Returning to the initial motif of the voluntary ghetto as the analytical framework used in this chapter to investigate the unique self-designation and socio-spatial pattern of transnational space within Tokyo, this last section attempts to summarize the findings and further embed the features of gaijin ghetto into the larger picture of the making of transnational spaces in the global city.

The term "gaijin ghetto" has turned out to be not only a simply deliberate and humorous wording used by transnational professionals, mocking their being disintegrated and discriminated gaijin in the Japanese society and self-ironically observing their spatial clustering within the urban space. There are indeed parallels to the scientifically studied phenomenon of the voluntary ghetto that apply to this particular socio-spatial pattern. Though, ghettos have been (and should be) designated areas of the dispossessed, the process of the ghettoization described by Wirth, that is:

"The immigrants drifted to the slum because here rents were lowest—a primary consideration. In addition, they found themselves within walking distance of their employment. Finally, here, in neighborhoods owned by absentee landlords, there was very little resistance to the invasion of people with a lower standard of living and an alien culture" (Wirth, 1969: 198), appears somewhat similar to the pattern of the gaijin ghetto, yet figuratively as a mirror inverted to the upper class. The transnational professionals have selected their housing according to their socio-economic means and needs (choosing the rents which were highest), finding themselves in perceived vicinity to their work and schools and, finally, feeling less confined in their social patterns and experiencing less resistance to their highest and Westernized "standard of living" and, in that sense, an "alien culture" to the local society. However, first and foremost, by the fact that transnational professionals are neither judicially, politically nor socially forced into these areas per se, the gaijin ghetto cannot actually be compared with ghettos of common understanding. The transnational professionals are the privileged upper class of society and are not as socio-historically discriminated against and dispossessed as the Jews in history. It must be emphasized that the gaijin ghetto is a terminology used by the professionals themselves and the socio-spatial seclusion still bears some resemblance to the conventional ghettoization in an inverted sense.

Yet, there are grounds to parallel the gaijin ghetto professionals to "[a] population [which] seeks an area in which it is tolerated and in which the wishes of its members can be gratified with the least amount of interference" (Wirth, 1969: 286). The spacing processes, such as the provision of multilingual businesses and services, and also signs and symbols found in the public space speak for such a tolerance and also the gratification "with the least amount of interference" as their everyday practices are being facilitated. Also, as Park notes in the foreword to Wirth's monograph, "the ghetto (...) owes its existence, not to legal enactment, but to the fact that it meets a need and performs a social function" (Park, 1969: viii). It is surely not a legal enactment in the case of these gaijin ghetto professionals, but it can be stated, too, that the gaijin ghetto is a socio-spatial structure that serves the social purpose of accommodating and concentrating a particular migrant group within Tokyo. The mechanism of the structural embeddedness of these migrants will be further discussed later, but already the example of the real estate agencies supporting the housing selection by mapping out the directions and bus lines to specific institutions located in the vicinity implies that the gaijin ghetto as a transnational space has its own social function.

And, further, as described by Wirth, too, on ghettos, "[o]nce in the area, each group tends to reproduce the culture to which it was accustomed in its old habitat as nearly as the new conditions will permit" (ibid. 283). This is another parallel which appears to be occurring in the gaijin ghetto when one considers the overlappings and limited expansion of social networks of the gaijin ghetto professionals during their life course changes. The reproducing culture of socializing also becomes apparent in the case of the professional who complains about the social ties within the gaijin ghetto society, where transnational professionals come and go while he remains in Tokyo. On the one hand, the social networks are actually kept over distance with the help of social media but also by the reproduction of the same ties in other locations to which corporate relocations bring the transnational professionals to. Conversely, the social practices are being reproduced in these transnational spaces of gaijin ghetto found in the network of such cities as Tokyo, since—according to the transnational professionals—these are the spaces where they are accustomed to the lifestyle and can keep their comfort zone.

Furthermore, taking into account these frequent comments on their feeling "comfortable" and an "acquainted lifestyle" within these areas of Tokyo, the synthesis processes of space-making can be reflected in the ghetto making, too. It is parallel to what Wirth describes as "the element of a common language, of community of ideas and interests, and the bare congeniality that arises even between strangers, who, coming from the same locality, meet in a strange environment" (Wirth, 1969: 20). As much imagined the common denominator might be, at least the same—or comparable—experiences of being high-skilled and mobile migrants appear to be the reason for the unique constitution of gaijin ghetto space among migrants from totally different ethnic backgrounds, perceiving the gaijin ghetto as their "comfort zone."

On the other hand, parallel to their "comfort zone" within the gaijin ghetto, the socio-spatial pattern out of this zone is also constrained by feelings of "uncomfort," too. Without language and intercultural skills, as one of the gaijin ghetto-type professionals account, his "*foreigner friends*" and he "*share the frustrations*" and feel uncomfortable around Japanese, where he cannot be as casual as with his foreign friend but feels "*like it's more formal*" (JH13). Such discomfort is also enhanced by the fact that migrants in Japan are mostly visibly outsiders among the predominantly Japanese or at least Asian society. Even a half-Japanese refers to being stared at on subway rides.

This then also brings us back to the conceptual issue of the (in)voluntary nature of these socio-spatial patterns of the "ghettoization." How voluntary or involuntary are the socio-spatial patterns of transnational professionals who are limited or focused on the gaijin ghetto as an area but also as a lifestyle?

The first and most obvious, yet crucial, determinant factor of transnational professionals' lives is their involvement in the transnational corporate context. The affiliation to the transnational corporate society pre-structures the professional in multiple dimensions. It is the work assignment and the occupation by which the overall living environment is embedded. It is first a corporate decision to relocate their professionals to specific locations, such as within the main hubs of the global cities network, whereas executive-level professionals are even involved in these corporate strategic decisions and responsibilities. Also, it must be noted that these corporate decisions are embedded in an ever-larger context of the global economy, which, however, will be discussed in more detail in the following chapter. This pre-selects the cities the professionals are placed in.

Apart from the locational issue, the corporate business also directs the duration of the assignments and sojourns of the professionals which, with regard to their socio-spatial patterns, heavily influence them, too. As we have seen previously, due to the usually shorter initial duration of migration around two to three years, with a prospective re-migration to another global city afterwards, the professionals are more prone to finding their comfort zones efficiently, which in the case of gaijin ghetto patterns is equated with consciously staying in the expats' area. As pointed out by some of the transnational professionals, the proximity to Western-style businesses is a convenience for the (perceived) limited time frame in which they relocate and with social ties to colleagues, international school friends and other people also located in these areas, such socio-spatial pattern even further sustained. The short initial duration, as one of the long-term professionals mentioned, also contributes to the stronger socialization of the same types of people, reproducing the effects of gaijin ghetto over time.

Time is also an impacting factor on transnational professionals' lifestyles and choices in the socio-spatial patterns, as efficiency in daily life appears to be a crucial aspect for professionals coping with the nature of their work in transnational corporations. As several professionals mentioned, their reasons for specific residential choices and also locational choices for leisure activities, such as after-work socialization, were the proximity to work as time is limited. The question of working hours was not further discussed nor mentioned by the transnational professionals. However, they often mentioned the shorter distance to work or quicker commute as an important aspect when deciding on the residential choice or even relocations within the city once they had arrived. Moreover, the lack of time for further leisure activities besides work and family life were mentioned several times. The mentions of these aspects emerge as clear indications of the time constraints caused by their work, which in turn impacts their socio-spatial patterns. With the lack of time also comes the lack of opportunity to meet people outside of the work or family contexts, which usually revolve around the children's schools and activities, so that socialization with people of similar circumstances accumulates.

Along with the time factor, the availability of capital is another crucial factor which impacts, if not decides, the socio-spatial patterns. It is not necessarily that all transnational professionals desire to live an exclusive lifestyle; however, the wish for comfort in such an objectively highly unstable work and living environment is indeed understandable. Some professionals also talk about the luxury that they wish to exchange for their work. In fact, the availability of monetary capital is a prerequisite to be living in areas of both gaijin ghetto and Pro-Tokyoite, socializing in specific clubs and schools and affording the usage of the upper-end stores and businesses in these areas. As will be discussed in the following chapter, the loss of monetary resources may have enormous impacts on the socio-spatial patterns of the transnational professionals and their families.

Closely connected to the monetary resources and also related to the corporate issue is the transnational corporate culture. Transnational corporate culture does not only decide on the overall acculturation to an interculturally sensitive communication and transnational working environment, which all interviewees confirmed as being accustomed to in their inter- and transnational corporate cultures. It also has the impact that affiliation to a certain global elite or an exquisite corporate elite results in social interactions with other—be it in formal or semi-formal business contacts or informal friendships developing out of the business—which again impacts the socio-spatial patterns. With regard to their identities, some transnational professionals even mentioned seeing the corporate identity or even industrial identity first before feeling attachments to ethnic or national identities. Such strong corporate and partly class identification surely contributes to socializing with very specific groups of people, both in the foreigners' and Japanese societies—both as an intrinsic desire to be socializing with the peer group and as an extrinsic motivation to belong to the perceived elite, and also as an external constraint as embedded in the overall work environment with the lack of time and occasions.

Another less obvious determinant factor for the transnational professionals' lives connected to the transnational corporations' strategies is the issue of residential choices made by the transnational professionals. Whereas transnational professionals do not lack the monetary resources for affording a preferred housing, the work environment does not necessarily allow a longer duration of searching for housings. At the same time, apart from the monetary resources, what these transnational migrants indeed lack are further social and cultural capital to cope with quick residential choices and also for societal integration. As the multiple migrations are part of their career, the usage of English as a lingua franca in corporate business is common and learning the local language for every single relocation is less common or even not feasible considering the time constraints and workload of their occupation. The business networks span worldwide, with which also some looser friendships establish through such connections. However, without having trusted social networks, such as family or kinship ties, as it more common with other transnational migrants, the resources that transnational professionals have access to in each of the locations are limited. When such business-friendship connections are indeed used, as those also tend to be socializing more in the corporate context than seeking socializations outside of work, the effect of the reproduction of the gaijin ghetto occurs as these professionals recommend within their realms of knowledge and networks.

One such phenomenon where several of these dimensions become clear is the quote from one of the transnational professionals who, in the corporate context of a prospective short-term sojourn and the time-distance to work and also under the constraints of resources, turned to a real estate agency as well as his corporate colleagues:

> My first thought was I assumed that I would live in Japan two years roughly, that was as far ahead as I thought. So I didn't want to live someplace where I had to spend a long time commuting, ehm so that's why I said, okay, all right, I am not doing that somewhere and that was where I spoke to the real estate agent about interviews to do a few places in Minato, Azabu, couple of places in my current neighborhood, some place near Gotanda and some places near residential area kind of near Shinagawa. (…) It was an agency that actually does handle a lot of expat type people, but they were pretty good, they spoke English which was pretty important for me at that time. (…) So, once I visited a few places and sort of tried to decide between … it was really between my current neighborhood and Meguro and I started asking my colleagues at that time, like, you know, these are the two places I am looking at, any comments, you know, just in general. (HU29)

As becomes clear in this quote, the other crucial part of the residential choice is the role of real estate and relocation agencies between the transnational corporate strategies and the residential choice of the transnational professionals. As one of the real estate agencies specializing in transnational professionals recounted, where in the past professionals received an extra business trip for searching for housing, nowadays with technological innovation relocations are primarily organized through internet communications without the professionals traveling abroad. A selection of a few housings, which match the needs and expectations of the foreigners—at least in the view of the real estate agents—are sent to the potential tenant and then contracts are organized through the Human Resource division of the transnational corporations. This procedure has changed even further after the global financial crisis as many transnational corporations have shifted from personal searches through local real estate agencies to outsourcing the task to relocation companies for even more efficient usage of resources. Another typical pattern would also be a longer-term relocation following a short-term assignment. In such cases, the transnational professional is initially accommodated in an upper-segment hotel, serviced apartments, or in short-term housing, which are found in the gaijin ghetto area, and then their relocation offered or decided. This has the consequence that, after being accustomed to the gaijin ghetto, the transnational professionals tend to be staying in the same area.[18] Such pre-selection and actually limited selection of the housing offers presented by the real estate agents—who maintain their full and professional understanding of the expectations of such expats—as well as the transnational corporate strategies that they are embedded in, predestines the socio-spatial patterns of newly arriving transnational professionals:

> As a foreigner, when you go to a real estate company and you say you're looking for a place, they have in their mind an ideal what you want. And

[18] In fact, to accommodate the needs of typical Western housing standards, such buildings are primarily built in the central areas discussed as being both gaijin ghetto and Pro-Tokyoite. Modern and larger housings with the amenities known from overseas have been, and are, built primarily in the newer building projects and also in the upper-class areas within Tokyo. As introduced in the chapter before, these relatively newer urban areas around Roppongi are where the foreign population, in particular the embassies and consulates, had historically been offered relocation from the Tsukiji Foreigners Settlement, and where a vibrant new urban development took place with lighthouse projects of the Mori Corp. so that English-language grocery and convenience stores, as well as the local businesses and other institutions, are in the vicinity providing a comfort zone for the transnational professionals.

> it's houses with, you know, maybe foreign appliances but different than the properties that they might show Japanese people with the same number of kids and the same whatever. Different properties, they show you altogether, certain properties, I'm sure, everybody who has rented that property for the history that it's been on the market has always been foreigners. (OP30)

The assumption voiced by one transnational professional about the different handling of foreign and Japanese customers by real estate agencies has been confirmed by the expert interviews with real estate agents, who were indeed having clear ideas of what not only foreign customers but also those highly skilled and highly affluent transnational professionals were searching for and desiring. Adequate client-orientation has been emphasized in the expert interviews:

> Well, these customers from overseas want their comfort; that is where we take them. As part of my personal services, as I need to have more than the competition, I take them to Costco and help them [in my private time] with shopping or the move (…) (Real Estate Agent 03)

which then also contributes to the drawing in newly arrived into the gaijin ghetto. Further accounts on the different ways of "accommodation the wishes" also showed that pre-assumption appears to be fairly strong with the agencies. The same agent described how he would take young and single male professionals to apartments with bath tubs with a skyline and imagining a nice night out and bringing back a date would be the right way to a deal for the rental.

While all mentioned the centrality and urban living as being the main residential choices of transnational professionals, interestingly, there were exceptions in these socio-spatial preferences; for example, with Germans, Canadians, Australians and Texans being mentioned as exceptions for their tendency to desire one-family houses with a garden instead of apartment living in the city center.[19]

[19] With their explanation based on their experience in the case of Germany due to the German school being located in the suburbs of Yokohama but also rarely having experience in, and therefore reluctance toward, urban living with "no real big cities there [in German]," and in the case of the latter three groups due to their wish for "barbecuing on their own lawn."

Though in many cases transnational professionals are indeed satisfied with what has been presented to them by these expert real estate agents, there are also cases being reported where venturing out of the gaijin ghetto for residential purposes was somewhat blocked by the real estate agents. Again, with limited resources regarding linguistic competence and also limited access to social networks as a newly arrived transnational professionals recounts the following situation:

> It was funny when I was looking for apartments I kept asking the brokers I do not want to be limited to just this thing, thirty-five choices here [in the Azabu area] and no choices outside. I said, what about Aoyama and Meguro and all these places I have heard great things about. We would like to look a little bit more than just here and I did not know the residential areas and I haven't been in them very much. So, I always felt like he kept driving in circles that I could not figure out where we were. And then I said 'Wait, that building looks familiar' and we make a ride and come back and then he showed me another apartment. I said, 'You are only showing me a part—like even if I ask you to go somewhere else.' (AB35)

This clearly shows the impact of strategies on the side of real estate agencies, in addition to the corporate context in which the transnational professionals are embedded. As it has been confirmed and discussed in further detail in interviews with real estate agencies, there is indeed a clear tendency to direct transnational professionals into specific areas *which are assumed (!)* to be comfortable and satisfying the needs and desires of these professionals. Indeed, as briefly mentioned in the previous chapters, there are differences in the interior architecture of traditional Japanese and Western-style housings, whereas in the upper-segment traditional Japanese housings are less common and less desired by foreign professionals. However, it must be noted that even explicit requests to go into Japanese upper-class housings are being dodged by the agents. There are several factors contributing to this phenomenon, such as purely economic reasons for the agencies as commissions depend on the rented facilities, the reluctance of Japanese landlords to rent out to foreigners (less common in upper-class housings but still existent) and also the cultural differences which might sometimes cause issues within the neighborhoods (e.g. throwing large receptions or home parties are uncommon in Japanese culture and, thus, may lead to neighborhood complaints). But these Pro-Tokyoites' requests of getting more immersed in the local society remain unheard.

In this context, it must be mentioned that transnational professionals are, though part of transnational migrants, distinct from other migrants as their first and foremost migration context lies in the corporate occupation, thus, migration paths depending primarily on the HR management and relocation companies chosen by the transnational corporations. Personal social networks, especially closer social ties, such as to other migrant family members or relatives, on which transnational migrant networks otherwise usually rely on, are irrelevant for these transnational professional migrants. The resources of the transnational migrant networks which usually are one of the main reasons and opportunities to migrate at all, but also available for the finding of housing, organizing everyday life and integrating into society, upon arrival, are virtually not existent for transnational professionals. Without having closer social ties to any locals or other migrants with local knowhow (apart from other professionally constrained colleagues) and the structural constraints of the HR management, the transnational professionals are dependent on the information of the transnational corporation and other resources available to them, such as English-speaking agencies.

The discussion of factors impacting and partly determining the socio-spatial patterns of transnational professionals sheds a new light on the issue of the voluntariness of the ghetto discussed in the previous chapter. Whereas from the account of the gaijin ghetto itself the image of a voluntary ghetto may have been evoked, since the motivation and active seeking of a familiar comfort zone appears strong, the overall work and social environment in which the transnational professionals are embedded in strongly questions the notion of voluntariness. In contrast to the original idea of the involuntary ghetto of Jews and other minorities who were not only judicially but also economically and socially put under constraints, of course, the transnational professionals remain totally privileged in their monetary resources and their socio-economic status and prestige. However, the voluntariness in the sense of having all the resources to freely select and choose their lifestyle does not seem to be existent either.

The transnational space-making of these financial professionals is not only extremely embedded in the transnational corporate structures and strategies but also constrained by this work and living environment. The lack of non-monetary resources, as well as the active but less obvious directions through the relocation companies and real estate agencies, further enhances the constraints for a truly voluntary socio-spatial patterning. Placed in-between the transnational corporate context and their individual characteristics and, thus, desires and preferences, transnational professionals' choices for socializing and finding locations for social interactions and social practices are more limited than it usually appears. Though it should not be exaggerated in the language, since these migrants are still socio-economically very privileged, it can be maintained that neither the socio-spatial patterns of the transnational professionals are fully voluntary. Especially if one looks at the distinct difference between gaijin ghetto and Pro-Tokyoite professionals, with the latter actively engaged in breaking out of the gaijin ghetto area, and if one follows the idea of the phenomenon as an inverted ghettoization, it can be maintained that the transnational professionals' particularly gaijin ghetto is a type of involuntary ghetto. Though there are no legal regulations limiting their residence elsewhere, there is a social constraint—both socio-culturally and through institutions like real estate agencies—confining them in their choices. The Pro-Tokyoites would then be those succeeding in venturing out and breaking out of the involuntary socio-spatial concentration in urban space.

The idea of the involuntary ghetto—yet inverted for a socio-economic elite—makes a comparison with other transnational migrants somewhat possible, whereas in terms of migration issues, these transnational migrants appear to be drawing only little from the migration networks for their migration itself, having only little or no social resources to rely on upon arrival. The networks that their migration experience base on are the institutional networks of being embedded in the transnational corporate contexts and having business networks rather than benefiting from actual individual social networks—at least when it comes to the migration and location in the arrival city. Furthermore, in the migration industry, in which they are particularly embedded, the relocation and real estate companies take a crucial role as mediators between the transnational spaces within the network of global cities. The insights into the socio-spatial patterns and the structures in which the ghettoizations of these transmigrants are embedded questions the notion of a privileged and voluntary choice into these areas which can be roughly paralleled to the idea of a non-place[20] (Augè, 1995), or typical "global expats areas."

References

Augè, M. (1995). *Non-places: Introduction to an anthropology of supermodernity* (J. Howe, Trans.). Verso. (Non-lieux).

Blakely, E. J., & Snyder, M. G. (1999). *Fortress America: Gated communities in the United States*. Brookings Institution Press Lincoln Institute of Land Policy.

Carter, H. (1977). Urban origins: A review. *Progress in Human Geography, 1*(1), 12–32. https://doi.org/10.1177/030913257700100103

Cybriwsky, R. A. (2011a). *Historical dictionary of Tokyo* (2nd ed.). Scarecrow Press. http://site.ebrary.com/lib/alltitles/docDetail.action?docID=10447164

Cybriwsky, R. A. (2011b). *Roppongi crossing: The demise of a Tokyo nightclub district and the reshaping of a global city*. University of Georgia Press.

Denton, N. A., & Massey, D. S. (1991). Patterns of neighborhood transition in a multiethnic world: US metropolitan areas, 1970–1980. *Demography, 28*(1), 41–63.

Fainstein, S. S. (2013). Assimilation and exclusion in US cities: The treatment of African-Americans and immigrants. In *Urban segregation and the welfare state* (pp. 28–44). Routledge.

Friedrichs, J. (1988). Makro- und mikrosoziologische Theorien der Segregation. In J. Friedrichs (Ed.), *Soziologische Stadtforschung* (pp. 56–77). Westdeutscher Verlag.

Friedrichs, J., & Triemer, S. (2008). *Gespaltene Städte?: Soziale und ethnische Segregation in deutschen Grossstädten*. (1. Aufl VS Verlag für Sozialwissenschaften/GWV Fachverlage.

---

[20] While the overall idea of the universal places which appear to have no own local characteristics appear sensible, of course, scientifically, particularly from a cultural geographical and spatial theoretical point of view, this anthropological concept needs to be scrutinized further. As these places, as unattractive it might appear to some people, were also attached specific meanings as spaces of encounters and shared experiences by the expats, these places do not necessarily become non-places. In fact, for them, these are crucial and valuable places in which they have their own socio-spatial practices with their peers. The question, thus, for a food for thought arises whether non-places exist at all or which other specific places could be more suitable for the phenomenon of a non-place.

Fusfeld, D. R., & Bates, T. M. (1984). *The political economy of the urban ghetto*. Southern Illinois University Press.
Government, T. M. *Statistics of Tokyo*. www.toukei.metro.tokyo.jp
Häussermann, H., & Siebel, W. (2004). *Stadtsoziologie: Eine Einführung*. Campus.
Hoffmann, C. (2012). Das Ghetto - eine Begriffs- und Diskursgeschichte. In F. Backhaus, G. Engel, G. Grebner, & R. Liberles (Eds.), *Frühneuzeitliche Ghettos in Europa im Vergleich* (pp. 53–77). Trafo.
Hollingshead, A. B. (1947). Selected characteristics of classes in a middle western community. *American Sociological Review, 12*(4), 385–395.
Lexikon der Geographie in vier Bänden. (2001). Spektrum Akademischer Verlag.
Lichtenberger, E. (1986). Stadtgeographie–Perspektiven. *Geographische Rundschau, 38*(7), 388–394.
Logan, J. R. (2013). The persistence of segregation in the 21st century metropolis. *City & Community, 12*(2), 160–168. https://doi.org/10.1111/cico.12021
Logan, J. R., Zhang, W., & Alba, R. D. (2002). Immigrant enclaves and ethnic communities in New York and Los Angeles. *American Sociological Review*, 299–322.
Löw, M. (2001). *Raumsoziologie (Suhrkamp Taschenbuch Wissenschaft)*. Suhrkamp. https://doi.org/citeulike-article-id:355221
Marcuse, P. (1998). Ethnische Enklaven und rassische Ghettos in der postfordistischen Stadt. In W. Heitmeyer, R. Dollase, & O. Backes (Eds.), *Die Krise der Städte* (1. Aufl, pp. 176–193). Suhrkamp.
Marcuse, P. (2005). Enclaves yes, ghettos no. In D. P. Varady (Ed.), *Desegregating the city: Ghettos, enclaves, and inequality* (pp. 15–30). SUNY Press.
Massey, D. (1985). New directions in space. In D. U. Gregory & John (Eds.), *Social relations and spatial structures* (pp. 9–19). Springer.
Massey, D. S., & Denton, N. A. (1993). *American apartheid: Segregation and the making of the underclass*. Harvard University Press.
Musterd, S. (2020). *Handbook of urban segregation*. Edward Elgar.
Musterd, S., & Ostendorf, W. (1998). *Urban segregation and the welfare state*. Routledge.
Park, R. E. (1969). Foreword. In L. Wirth, *The ghetto* (pp. vii-ix). University of Chicago.
Peach, C. (2005). Social integration and social mobility: Spatial segregation and intermarriage of the Caribbean population in Britain. *Ethnicity, Social Mobility and Public Policy*, 178–203.

Ravid, B. (2012). On the diffusion of the word "Ghetto" and its ambiguous usages, and a suggested Definition. In F. Backhaus, G. Engel, G. Grebner, & R. Liberles (Eds.), *Frühneuzeitliche Ghettos in Europa im Vergleich* (pp. 15–38). Trafo.

Robinson, J. (2006). *Ordinary cities: Between modernity and development* (Vol. 4). Routledge.

TAC. *Tokyo American Club - About us*. https://www.tokyoamericanclub.org/index.php/en/about-the-club

Van Kempen, R., & Murie, A. (2009). The new divided city: Changing patterns in European cities. *Tijdschrift voor Economische en Sociale Geografie, 100*(4), 377–398.

Van Kempen, R., & Şule Özüekren, A. (1998). Ethnic segregation in cities: New forms and explanations in a dynamic world. *Urban Studies, 35*(10), 1631–1656.

Varady, D. P. (2005). *Desegregating the city: Ghettos, enclaves, and inequality*. SUNY Press.

Wacquant, L. (1994). Morning in America, dusk in the dark Ghetto the new «Civil War» in the American city. *Revue française d'ètudes amèricaines, 60*, 201–206.

Wacquant, L. (1997). Three pernicious premises in the study of the American ghetto. *International Journal of Urban and Regional Research, 21*(2), 341–353.

Wahsaw. (2019). Tokyo subway system map Japanese. https://commons.wikimedia.org/wiki/File:Tokyo_subway_system_map_Japanese.png

Wilson, W. J. (1993). *The ghetto underclass*. Sage.

Wirth, L. (1969). *The ghetto* (10th ed.). University of Chicago Press.

5

Avoiding the Gaijin Ghetto: Localizing Pro-Tokyoites

> Well, I didn't wanna live in a expaty kind of ghetto. You know, I didn't want to live somewhere like that. And I just liked the area. It's a beautiful area and it's more local, you know. Don't get me wrong, it's a high neighborhood, like three movies stars in my building, you know, but they were all Japanese, right? (JN17)

Whereas the previous chapter discussed the gaijin ghetto of transnational professionals, which is characterized spatially by its very dense and limited extension within Tokyo and socio-spatially by its multilingual (mainly English-speaking) global character, this chapter deals with another phenomenon going beyond the gaijin ghetto—beyond both spatially and socially. As hinted by the quote above, there is a group of transnational professionals who actively and consciously avoid the gaijin ghetto and venture into what is, or is at least perceived as being, more local to the Japanese society. At the same time, becoming more local is, as also implied above, not to be equated with adapting to the average population of Japan but considered to be keeping a lifestyle which is different from the gaijin ghetto and yet still "high" in socio-economic and socio-cultural terms. In other words, the endeavor behind is to be global expats

S. Yamamura, *Spatial Diversity in the Global City*, Global Diversities,
https://doi.org/10.1007/978-3-031-64725-3_5

who share spaces with the local upper or privileged class, becoming their global-local peers.

The main aspect of this chapter is to further delve into the question about what the socio-spatial patterns of these people are, who do not identify and comply with the gaijin ghetto and especially what makes them different in their processes of space-making while also discussing factors that may impact the different patterns. As it becomes clear, they overlap with the gaijin ghetto but also begin to venture out of their spaces. Thus, the chapter will analyze what make these patterns distinct in the socio-spatial pattern as well as in their social practices and the synthesis process to gaijin ghetto patterns. It will also be further elaborated what the grounds are to be designating these patterns "Pro-Tokyoites."

5.1 Avoiding the Gaijin Ghetto as a Type of Transnational Attitude

Starting off with the socio-cultural attitude of the category of transnational professionals—henceforth named Pro-Tokyoites (an explanation on this designation will follow)—what is distinct is the active decision to venture out of the gaijin ghetto. They explicitly talk about a "*conscious effort*" (BM21) and "*conscious choice*" (CD23) to avoid or leave the gaijin ghetto. Also, apart from the voluntary choice, the distinction from the gaijin ghetto category appears to play a role in their consciousness, too, as phrases like "*[W]e go much, much further. We run further*" (IN76) or the active usage of the personal pronoun "they" in contrast to the "we" and "us" could be observed.[1] The main reason for this pattern lies in the intrinsic interest in the local society and culture, rather than the

[1] While such differentiation might appear as an onset of *othering* mechanism, the distinctions to the others in plural are a more complex issue here. In this particular case Pro-Tokyoite would consider themselves somewhat different from the gaijin ghetto patterns, but would otherwise perceive themselves for sure part of the overall expats community. The differentiation to non-expat foreigners or non-highly skilled professionals, such as language teachers, let alone to lower-skilled labor migrants is far more stronger as well as to the local average Japanese population to the fine distinctions within the expat population.

preference for comfort expressed by staying in the otherwise accustomed environment of the gaijin ghetto:

> Ok. That was, that was a conscious choice on my part, because when I arrived there back in '88 everyone was living in what was called the gaijin ghetto, which was over near Hiro'o. And I said: you know, what's the sense of moving to Asia and living with a bunch of foreigners? I'd rather move into a neighborhood where, you know, there are not so many foreigners perhaps. (CD23)

Unknown and, at first, estranging cultural traits of the local neighborhood are perceived as a positive experience, if not an exciting one, compared to the well-known Western style of living. As the same interviewee narrates on his first unusual experience of a Japanese neighborhood custom on fire prevention,[2] which he encountered when he first came to Tokyo in the late 1980s and settled into a non-gaijin ghetto area within the city:

> And, you know, it had a nice feel to it. The guy came through the streets at night, banging the wooding sticks, reminding everybody to turn out the fires. And I was like: what's that guy doing? And somebody told me and I was like: that's wonderful! So it was some of that old culture still in that neighborhood, (…). (BM17)

As he continues with the account of the local neighborhood he has chosen to live in, local businesses are also mentioned as a positive aspect which can be found there. Whereas the gaijin ghetto is characterized by English-speaking businesses and imported goods, though partly owned by Japanese entrepreneurs, too, the more local spaces are characterized as being mostly Japanese in the residential population but also in the urban economy. As he recounts:

[2] In the wintertime, members of the neighborhood associations patrol through the neighborhood making warnings with resonant woods to remind of the danger of fire. This is indeed an old custom which derives from the wooden architecture in Japanese housing and has recently become quite rare, especially in metropolitan areas like central Tokyo.

> …and the local shops were still there. You know, the place where I took my shirts to be washed and ironed was a little husband-wife operation in this very small, old wooden building. Everything was done by hand, which I just absolutely loved. So, I made a conscious effort not to, if you will, move into these very Western neighborhoods. (BM17)

In fact, it must be noted that such family businesses or other "mom-and-pop stores" have become generally rather rare in Japan (and also globally) compared to the late 1980s when the interviewee first moved to Tokyo. However, the same affinity to local Japanese businesses is also prevalent in transnational professionals of the later decades in the late 2010s. The transnational professional who has moved to Tokyo recently mentions:

> We also say that we, where we shop for food varies … we go to Japanese supermarkets when, often because ingredients we can find were better or more interesting and we were willing to […] experiment a lot with that sort of thing. But there are definitely times where we feel like we need to just take it easy and we do not want to challenge… we can go to the National Azabu where everything is products from around the world that is very easy, everybody speaks English. So there is definitely a […] it is easier to feel safe and when you do not feel like making the effort of really being in a place where you do not speak the language, you can make it easy. (AB34)

What additionally becomes clear in the previous quote is that language skills remain a barrier, or also a key, to immersing into the local society. The transnational professionals are—though willing to—restricted in their choices and behaviors by the foreign language. However, what is distinct for these professionals is not their proficiency of Japanese or the lack of it, but the willingness and interest in confronting themselves with the unknown. As mentioned before, mastering the language was not necessarily a prerequisite nor a criterion for exclusion to showing a stronger gaijin ghetto pattern—there were in fact also people who were fluent in Japanese or even had a Japanese spouse yet primarily remained in the gaijin ghetto. Likewise, venturing into the more local areas of the Pro-Tokyoites is not necessarily coupled with language skills, either.

Furthermore, an important aspect of the Pro-Tokyoites which is illustrated in the latter quote is also the alternating usage of the Japanese and global stores and spaces. This is an important aspect to point out as the Pro-Tokyoite mindset is not what is commonly associated with the idea of "going native," by drastically leaving the foreign lifestyle and living "purely Japanese."[3] This would not only be impossible, or at least very difficult, to comply with in their professions at transnational corporations, especially in the financial industry which is by nature globally driven and English-speaking but also requires high intercultural skills and mobility for business trips and overseas assignments. It would also not be the aim of these professionals as the pattern is a "venturing into" local spaces in addition to their more well-known comfort zones and not fully leaving the gaijin ghetto. It is this flexibility and openness that is characteristic of the Pro-Tokyoites pattern, which can also be observed in their socio-spatial behavior elsewhere in the world wherever they migrate to.[4] The following quote summarizes the attitude of these transmigrants:

> I'm pretty relaxed that way, so there's an example, I like to go, I like to see a variety and being local at the same time I don't want to be a burden. (…) I try to get like a local-centric, something unique. (RL09)

5.2 Pro-Tokyoite Socialization with Gaijin Ghetto and Local People

This attitude to live in both a global and a local environment is reflected also in their social interactions, which are characterized by an overlapping with the gaijin ghetto on the one hand and an extension of the social networks beyond them on the other. Depending also on the spouse—yet not necessarily the language or ethnic background of the person but rather the attitude, too—the pattern of socialization differs according to the previously presented attitude:

---

[3] In times of globalization, it is questionable what exactly "purely Japanese" means nowadays, but what is meant here is the pattern of foreigners who not only find Japanese friends but also even more, for example, by moving out to the countryside to aim for experiencing the rural life, and become acquainted with customs of Japanese society which have even become rare and strange to the urban population, especially the younger generation.

[4] The extent of the global phenomenon will be further discussed in Chap. 7.

> My perspective on that is, why go somewhere new, if you only gonna hang out with people like you … or only do things with people like you, right? The purpose of going somewhere new is to… I think a huge willingness to want to explore and want to integrate. I met people that only want to go to the Western grocery stores, to the Western schools and interact with other people like them. But, for me that takes the value away from an international experience. (KW44)

Being involved with the "other" professionals who stay mainly in the gaijin ghetto through the work environment and the more or less obligatory accompanying semi-formal encounters of after-work events—as one interviewee pointed out: "*Actually, all my friendships are based on relationships made through work. That was my only way to meet people in the beginning and they all stem from that*" (AC07)—these Pro-Tokyoites who venture into the local spaces with their families also do so regarding their social networks. When asked about their social interactions, these transnational professionals mention befriending Japanese colleagues of the same transnational corporations or colleagues with Japanese spouses who would be mediating the social integration in Tokyo.

For such communication and involvement with Japanese society, the language skills—especially knowledge of the local culture and customs—can be crucial, whereby the role of the Japanese spouse intuitively appears to play a decisive role. In fact, the Japanese spouse can be a reason to be integrating more into Japanese society, especially through the family and friends of the spouse. The Japanese spouse can be contributing to venturing out of the gaijin ghetto by co-deciding on the residential choices, for example, "*My wife does not want to live near the school because of horrible international school manners*" (IN20), or providing the option to use non-English-speaking businesses: "*So she goes there and some supermarkets on the way of our area if we need things. But yeah, pretty much local*" (PL10). It is common that those with a Japanese spouse tend to also have some connections to locations on the outskirts of Tokyo if the spouse's family lives there. The personal relationship and with that a higher level of language skills or even the frequency and duration of stay in Japan can strongly impact the localization of a transnational professional.

However, interestingly, there are also several cases where transnational professionals mention that their integration is limited as the Japanese spouses' social networks are quite focused on the gaijin ghetto or even their preferences have contributed to their moving into the gaijin ghetto area: "*My wife certainly said, you know, it has one of the best reputations even among the Japanese people. That Ebisu is a very, very good neighborhood*" (IG04) or using Western-style gastronomy because "*often my wife's friends are looking for an excuse to go to some place a little more Western*" (AC15). There is also even an account of an active avoidance of the Japanese community by the Japanese wife or the limited involvement in the local community:

> My wife is very aggressive. I think she had lived in Deutschland in junior high school and went to the UK for high school, that' s why she has some kind of... So she avoided to communicate with the Japanese, she tried to communicate with local people. (INT32)

The impact of having a Japanese spouse should not be underestimated, but as the few examples of Japanese spouses previously also illustrate, the mere proficiency in a language and/or the marriage to a Japanese spouse does not automatically result in a detachment from the gaijin ghetto. As a matter of fact, even non-Japanese spouses can contribute to venturing out of the gaijin ghetto—dependent on the individual mindsets of the persons involved. The following quote exemplifies the gradual integration of a transnational professional's (non-Japanese) wife into Japanese society, whereas venturing out of the gaijin ghetto pattern is described as a mixture of a change that the person has been aware of, yet less consciously deciding on:

> [I]t is funny because I was having this talk with my wife the other day about her relationships involving friendships with people over the last year. At first it was closer with a lot of the people that were sort of similar; both Western spouses, both coming over here on job assignment or something like that. And how she has migrated this year to [...] probably spending most of her time with Japanese women who happened to be married to Westerners (...) It is an interesting observation that she had about working

> here and the people she is interacting with and the types of activities that she is doing or places she is going. And it is certainly a lot of crossover but it is also something that she has been aware of and it was not a conscious decision or anything. (AB010)

Again, language skills are important for opening up possibilities, but what is more significant in the end appears to be the lack of fear of contact and thus the cultural flexibility. Though their skills (spouse is also non-Japanese) in the language are limited, the transnational professional explains how he receives and actively asks for help by local colleagues and is open for recommendations from locals, and also does not hesitate to visit restaurants which require Japanese language skills:

> And a lot of our friends are mixed couples where there is a Japanese spouse and a Western spouse. So we do get a lot of chances to do more sort of Japanese activities whether on weekends or travelling or restaurants or whatever with our friends and their spouses because… they speak Japanese and can help us do things we would not always be able to do on our own successfully. (AB010)

What is also mentioned in this context of not mastering the language but still finding a way through non-English-speaking environments is the meaning of technology. Whereas gaijin ghetto professionals tend to refer to the technical innovation and development of internet and communication devices solely as a support to maintain their transnational practices by Skyping and being directly involved in the home society's daily events, Pro-Tokyoites see the same digital devices additionally as an important tool to immerse themselves into the local society overseas:

> And certainly, I think a big help is frankly the iPhone and technologies that […] you can always find a map and sort of show somebody where you want to go in the worst-case scenario. We have translation features on the phone that you can speak into and Google translate field. You can do a Google search on any town and if I cannot find the place you are looking for then you know at least how to get back to the train station or the hotel or something. (AB10)

Though duration of the stay plays an important role in any socio-cultural integration and thus to the distinct Pro-Tokyoite versus gaijin ghetto patterns, as well as the language skills, but as already explained in the previous section, there are not only a few transnational professionals who remain in the gaijin ghetto during their long residence and despite their Japanese language skills. The following quotes illustrate that the short duration or the language skills are not an essential criteria for sticking to the gaijin ghetto or to the "*expat bubble*" (KW44), as both transnational professionals have come from New York to Tokyo only recently, less than 1.5 years ago at the time of the interview, and already show very different patterns with regard to the similar situation of dining out:

> We have a lot of help with searching for things. A lot of it comes from recommendations, from Japanese spouses of friends, that were friends with these couples. Some of them from ... come from clients [who] recommend areas or specific places. Certainly, the entire staff here speaks Japanese and, you know, is able to help make reservations in an easier way. So, if I call the ryokan[5] hotel, I say: please tell them that we do not speak Japanese so don't ask us many questions while we are there unless [laughs] they speak English. We go to restaurants and again, well, we don't know whether they speak English or don't before we go or, if they don't have English menu, we'll just learn how to say: 'Osusume wa nandesuka'[6] very quickly, so that we can usually get fed anywhere that we go. (AB_10)

> In terms of, you know, my foreigner friends, we all sometimes have the same experience when we go to a restaurant, share the frustrations as much as we try to speak Japanese, it's maybe our accent or pronunciation, the other person doesn't understand it, but we know that we are saying the word, we just can't make it happen, but a very similar experience (...) You know, when we ... socialize with my non-Japanese friends. I think, it can be a little bit more casual where it's hard to offend? When I am around more with Japanese friends, I feel like it's more formal. (JH13)

[5] Ryokan (Japanese): traditional Japanese accommodation which typically features tatami (rice straw)-matted rooms and communal (hot spring) baths and compared to other types of accommodation is usually more expensive.

[6] "Osusume wa nandesuka" (Japanese): translates to "What is the recommendation?"

Whereas the first professional has no hesitation, or if so only little, to interact with Japanese and actively search for recommendations and assistance by Japanese clients, colleagues and acquaintances, the latter shows stronger reluctance to socializing with the Japanese. The same experience of being "lost in translation" in a restaurant is perceived and coped with completely differently. The first does not see a problem in not being able to speak Japanese but finds a way to cope well with the situation, including the proactive learning of Japanese phrases to get by. The other professional, in contrast, reports feeling frustrated by the situation and then somewhat retreats to the comfort or convenient zone of the gaijin ghetto where he has other professionals with whom he can share the feeling of discomfort outside in the local society.

Despite the relativization of the importance of language skills, there is no denying that fluency does have an impact on the abilities to both socialize with locals and to use local businesses. Yet, taking a closer look at those professionals who were also Japanese citizens or had some Japanese background (e.g. children of mixed ethnic backgrounds with one Japanese parent), it can be maintained that the socio-spatial patterns tend to differ irrespective of the Japanese language skills, which they all had.

Whereas one "half"-Japanese (i.e. binational or biracial Japanese, often with dual citizenship or at least a dual ethnic background; in this case someone with strong Japanese language and intercultural skills, and also extensive experience abroad by growing up in the UK and migrating through work to Singapore and Tokyo, for example) showed strong gaijin ghetto patterns with many foreign friends to socialize with in the Roppongi area:

> Yeah, I go [to Roppongi] with my friends as well, but my friends are typically foreigners as well. Well, I hand around with both, Japanese and foreigners. But we do go to Roppongi quite a lot. (…) I think being in the industry, there are just so many foreigners working… I just tend… And they're usually the ones who tend to go out and drink and have a good time. I think that's a fair statement! [laughing] I don't know, I just naturally seem to click with them more than the Japanese people. (ML01)

Another "half"-Japanese (also one Japanese parent and one with a different ethnic background on the other side, with similar diverse and extensive experiences outside of Japan) explained to have a mixed background of friends, with both foreigners and Japanese and also by going to different locations:

> Everybody goes there or used to go there. I was the first few years in Roppongi, so Harpland, I guess it's very foreign, it's a standard bar, you don't see any of those in Japan. (…) In Roppongi you would tend to go to the Western influence bars. You don't sit down, you have a drink; you don't have a meal, you have a snack and you drink and everybody's standing at the bar. There are a couple of bars like that in Roppongi and people would just do hashigo[7] from every single one of those. So, you would stay clear from the Izakaya.[8] That's where you go out with your work friends including lots of Gaijin. (…) It's a mixed group. Japanese people, there are some who enjoy being in a company with Gaijin, it's funny, it's fun, they would follow or stick together and go around with the group. International school people I met, ehm, they're totally mixed. They would sometimes go to an Izakaya, it's a … yeah, Western bars, clubs are always slightly Western. (AC07)

Here again, it becomes clear that merely a Japanese background of people to socialize with does not automatically result in venturing out of the gaijin ghetto. Depending on the Japanese people themselves, for example having an "international school" background, a transnational professional could be socializing both in the gaijin ghetto and in more local places, such as the Japanese pubs (izakaya).

What the previous quotes imply is the diversity of sub-types within the pattern of these Pro-Tokyoites, which derives from the diversity of the relations to Japanese language and culture. As transnational professionals also encompass, as mentioned in the methods section before, Japanese citizens or those with a Japanese ethnic background and extensive experience abroad through their profession in transnational corporations, there are subtle differences within the patterns of these persons. Japanese

[7] Hashigo (code-switching into Japanese): Japanese word for ladder, colloquialism for bar-hopping or pub crawl.

[8] Izakaya: Japanese-style pub in contrast to Western-style bars or clubs.

citizens who have grown up in Japan, or even studied in Japan before going abroad, would show stronger local characteristics in their socio-spatial patterns because they would also be more embedded in the Japanese society. At the same time, half-Japanese professionals, or those growing up abroad and "returning" as students or adults, would be more integrated into a society indicated previously as "international school people," which consist of either Japanese brought up abroad (*kaigai-shijo*) or returning as a youth from abroad for educational purposes (*kikoku-shijo*) or even Japanese fully growing up in Japan but being educated in English-speaking international schools. Half-Japanese transnational professionals tend to socialize better with them than with the "purely Japanese" (see previous). However, interestingly, the interviewee who discussed this issue also pointed out that he—as a kaigai-shijo but also dual citizen with dual ethnic background—does not feel attached to other Half-Japanese[9] but rather to these "*international school people*" (AC07).

5.3 Socio-spatial Implications of Pro-Tokyoites

Continuing with the spatial implications of the Pro-Tokyoite pattern, the difference between them and the gaijin ghetto patterns observable in the mindset and the socialization is quite apparently reflected also in the spatial dimension. In a direct comparison that follows, it stands out that the locations mentioned by these transnational professions overlap with the

[9] "*There are communities. Depending on what you are looking for, there are a lot of things. There is even a group of half-Japanese people. I can't remember what they're called, but 'Happa Japan'? I think. I went to couple of their events, I didn't really feel comfortable to be honest, it wasn't me. It was all, I guess the problem was, the only thing we had in common was we were all half-Japanese, so I found a… I still feel that half-Japanese people sometimes fall into, have these very strange insecurities that I've never felt myself, to be honest. They say they don't belong to either culture. There is even a website these days I found, which is dedicated to half-Japanese people saying that they don't belong to either culture, they aren't recognized in either country, so they kinda feel left out, insecure, have nowhere to go. They don't have an identity, what they say. I don't feel that at all. I think it's quite mysterious and I guess, it's something you see a lot in people with mixed backgrounds, but personally I don't relate to that much*" (AC_007). The issue on identity, sense of belonging and also community is indeed an interesting aspect which emerged during the course of research, but as this would go beyond the scope of the study and is not closely related to the socio-spatial phenomenon studied, this will not be further discussed here.

gaijin ghetto as already mentioned in the previous section, but also spread beyond the golden triangle and shows stronger diversification (more locations mentioned and the isolated areas on the outskirts of Tokyo Prefecture) (Fig. 5.1). Moreover, the centrality of the Pro-Tokyoite space within Tokyo also stands out, which is an important criterion for their designation as being Pro-Tokyoites.[10]

An interesting phenomenon in this context is the image of "venturing out" of the gaijin ghetto and the dispersion of the socio-spatial patterns, which is reflected in their mental maps. The imagined borders of the triangle or the gaijin ghetto, which had been drawn in by most gaijin ghetto-patterned transnational professionals, are not observable and—apart from orientation borders of the Bay of Tokyo and the Imperial Palace often drawn in as a circle—strict borders as in the case of the gaijin ghetto are not drawn (Fig. 5.2). This is significant in that the perceptive, reminiscent and imaginative synthesis process of space-making is directly expressed in these mental maps of Pro-Tokyoites in contrast to the gaijin ghetto. The venturing out is often also visible by arrows leading out of the central area of Tokyo (Fig. 5.3). The indication of transportation means, that is, the subway, is still often an element to be found.

Concrete areas, which these transnational professionals mention, are the surrounding areas from the Golden Triangle of the gaijin ghetto. As one professional made clear:

> So, and if we go to a restaurant and the guy … sometimes some of them do not know their way around Tokyo so the easiest one is do something close to the hotel so the guy can go because it is their hotel and they can get home. So, chicken and eggs whether you are going to Roppongi, Akasaka, Nishi-Azabu, Aoyama. It is still very much this sort of gaijin ghetto, I guess. The guys who lived in Tokyo a lot longer and still living here we go much, much further. We run further, [Laughs] so occasionally

[10] Individuals also mentioned some locations in the outskirts of Tokyo or adjacent prefectures, such as the place of residence of the Japanese spouse's parents or other individual-specific areas, but as they are more based on individual biographies and do not reflect the overall pattern of this group of people, they are omitted from the map.

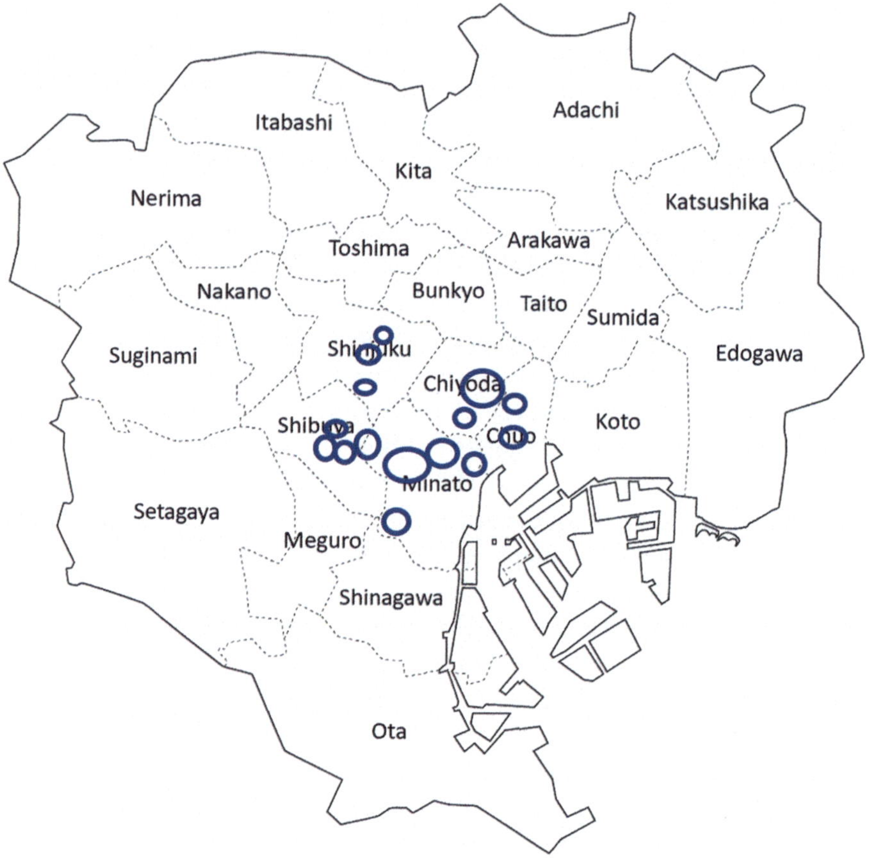

Fig. 5.1 Pro-Tokyoite transnational space in Tokyo (author' own)

Shimokitazawa[11] or kind of like natsukashii.[12] I remember I used to live out here (city center) or Shibuya or Shinjuku or … It is pretty boring really. Probably as far as Shinjuku, Kagurazaka, maybe then sort of kind of

---

[11] Shimokitazawa mentioned here had not been listed as a socialization area with other Pro-Tokyoite professionals, except for those who had been in Tokyo as younger non-professionals before—which is also the case with this transnational professional who had spent additional time in Tokyo in the past as a student—and tends to be a hip area for the youth and students.

[12] Natsukashii (code-switching): Japanese phrase literally standing for "dear, beloved, cherished" or "missed, longed for"—used for expression when reminiscing fondly about some past events, situation or object.

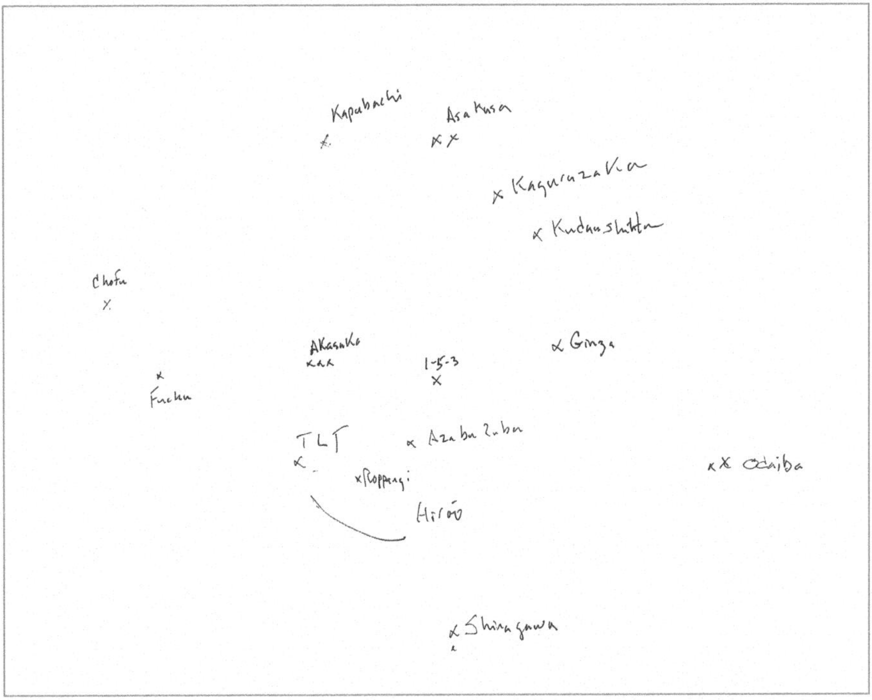

Fig. 5.2 Mental map of Pro-Tokyoite (1) (research material of author)

> Nihonbashi, occasionally because the Mandarin Oriental is over there and there are some good restaurants and over there a few other and then you come sort of the higher society including here to Ginza, Shiodome. (BM37)

Adding a few more locations, such as Shirokanedai and Daikanyama—places also frequently mentioned by Pro-Tokyoites—to the picture and juxtaposing the maps of Pro-Tokyoite and gaijin ghetto patterns, it becomes very clear that the spatial extension of the Pro-Tokyoites pattern is, in accordance with their image of breaking out or "running further" and also overlapping social practices with the gaijin ghetto, encompassing the gaijin ghetto area and also stretches further north, south and east. Areas in the east, such as Ginza, Nihonbashi and Shinbashi are interestingly perceived by gaijin ghetto transnational professionals as being either touristy or even too Japanese:

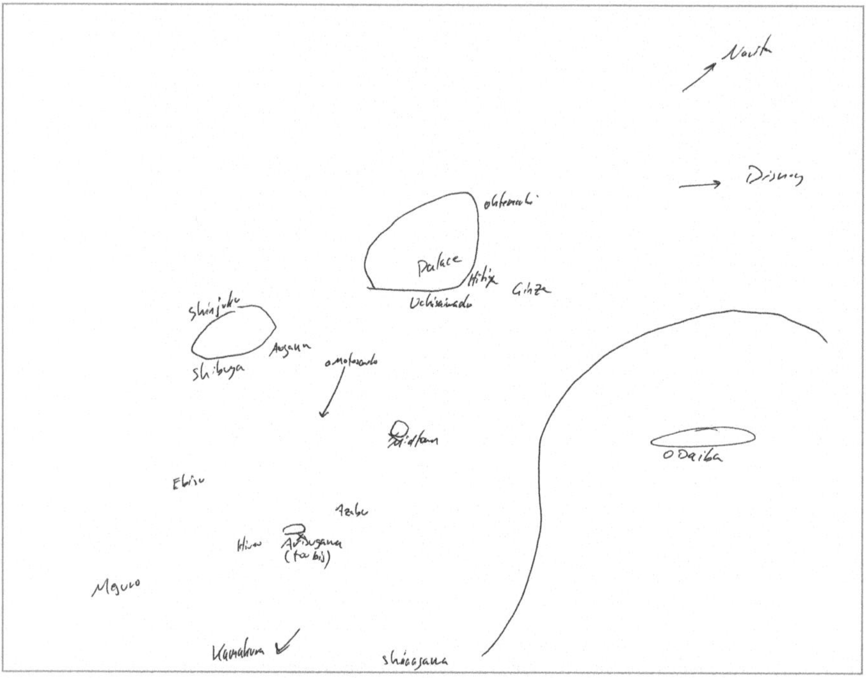

Fig. 5.3 Mental map of Pro-Tokyoite (2) (research material of author)

> I don't know, ehm, you don't see many foreigners out in Ginza, maybe. You have the tourists obviously, but just to see the sights and the shops. All the restaurants, Izakaya, lots of Japanese people. It's very Japanese in Ginza. Same as Shinbashi. (LP07)

The most interesting aspect of the Pro-Tokyoite space is, as alluded to previously, that when this area is mapped out as the Pro-Tokyoite map (with the Golden Triangle of the gaijin ghetto included as a reference) it basically corresponds to the socio-spatial patterns of the local peer group, that is, Japanese upper-class Tokyoites (Fig. 5.4). The local peers were defined as those Japanese who have the same socio-economic and educational background (highly skilled), and also being natives to Pro-Tokyoite (in all cases, at least one parent was a Pro-Tokyoite already and all of the

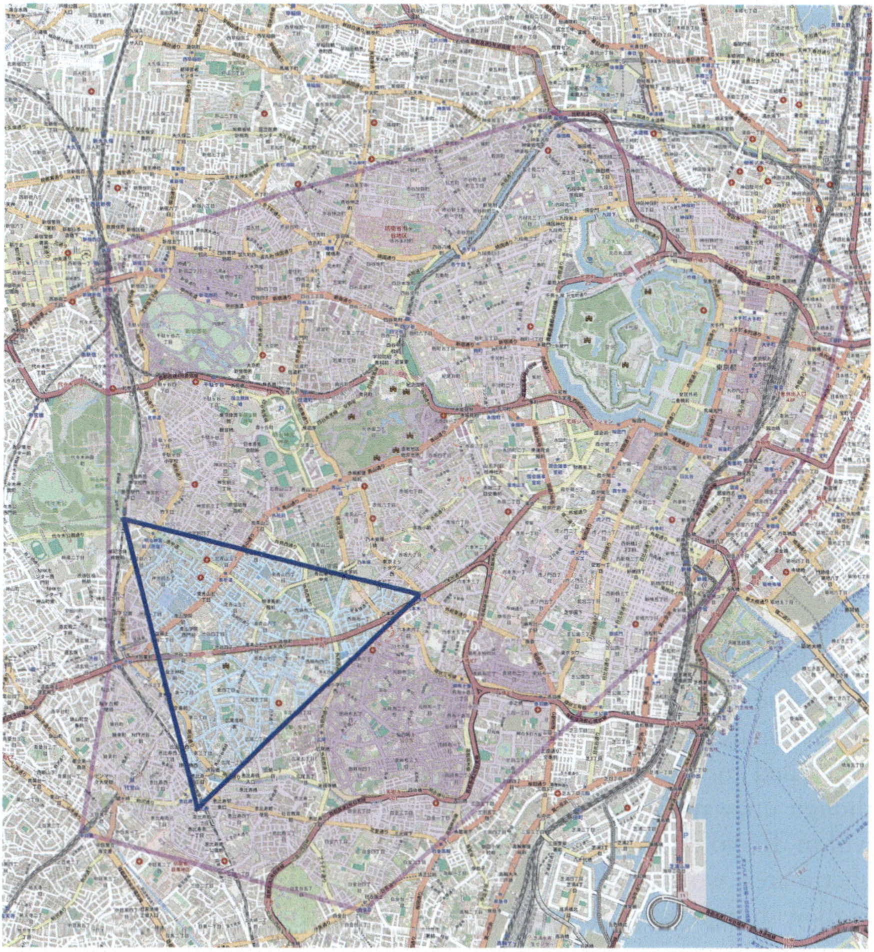

Fig. 5.4 Pro-Tokyoite area in Tokyo in relation to gaijin ghetto golden triangle (author's own, based on OpenStreetMap)

group interviewees were born and raised in Tokyo in upper-class families). This is how the naming of "Pro-Tokyoites" derives from.

As part of the triangulation, a group discussion with elderly, upper-class Japanese Tokyoites was carried out to deduce whether the socio-spatial pattern of this particular group of transnational professionals was in any way similar to that of locals. Moreover, the real Tokyoites actually also avoided the gaijin ghetto area, indirectly criticizing[13] it as being a new area for young people as well as where many foreigners hung out. As it can be seen from statistics, the area of the gaijin ghetto has indeed a young population and their rather dismissive mentioning of this area derives from the fact that the area around Roppongi was historically negatively associated with the GI culture[14] in the post-war period as a "gaijin" area, and also the background as a place where there was basically nothing when these Tokyoites grew up apart from "bushes and woods." As a matter of fact, the older generation of upper-class Tokyoites would not go beyond the borders of the Yamanote area, which was historically already the upper-class area south and west of the Imperial Palace. The areas of the eastern part of central Tokyo were especially dismissed by them as the "Edokko" and Shitamachi area, where they would have no knowledge about the whereabouts. The Shitamachi ("under city") would traditionally be the location of merchants and where one would still find traditional handicraft and residential areas of the working class.[15] At the same time, the area within the Yamanote railway line—but not the actual Yamanote area, that is, the northern part from the Imperial Palace—would also not be used or even mentioned by the Tokyoites.

On the basis of this similar socio-spatial pattern of the transnational professionals to the local upper class, the designation of Pro-Tokyoites developed for them. The interesting issue is that these Pro-Tokyoites, though unaware of the historical background of the socio-spatial divide within Tokyo and ignorant also of the Tokyoites' notions of the gaijin ghetto area, would still follow the socio-spatial patterns of the Tokyoites as they, though being foreign or non-Tokyoites themselves and avoiding

---

[13] It should be noted that the group discussion was quite difficult in the analysis as the upper-class Tokyoites would not be direct in criticizing or degrading people or specific areas, but only allude to the notions. The interpretation has been, however, cross-checked with a closer informant who had also participated in the discussion and would agree to be clear in the wordings in the follow-up face-to-face interview.

[14] Roppongi has been a red-light district that flourished during the time of the US occupation of Japan in the post-war period, and only recently developed into a new urban hotspot for upper-class businesses and buildings (see more in Cybriwsky, 2011); the Tokyoites perceive the Japanese population interested in moving to these areas as also being nouveaux-riches, who are 'wannabe-Tokyoites'.

[15] cf. Seidensticker (1991): High City, Low City.

the gaijin ghetto, still remain in the socio-economic patterns of their peers. They are indeed Pro-Tokyoites, if not Yamanote-Pro-Tokyoites. This is why the introductory quote not only perfectly illustrates the venturing out of the gaijin ghetto, but also explicitly points out the keeping of the standard, the "high neighborhood" with Japanese movie stars.

With the previous chapter on the gaijin ghetto and the present on the Pro-Tokyoites, two somewhat extreme cases on the broad spectrum of socio-spatial patterns within the migrant population of transnational professionals have been presented. As these patterns almost erroneously confer the image of transnational spaces as being rather static, the next chapter presents the third empirical result of the research, which shed further light on the socio-spatial diversity in global cities. Introducing the third pattern of transitionists and also contextualizing these presented socio-spatial patterns into the dynamics of temporal-spatial issues, the following chapters opens the debate on the actual nature of spatial super-diversification as a process and breaks beyond the notion of the superdiversity.

References

Cybriwsky, R. A. (2011). *Roppongi crossing: The demise of a Tokyo nightclub district and the reshaping of a global city*. University of Georgia Press.

Seidensticker, E. (1991). *Low city, high city: Tokyo from Edo to the earthquake: How the Shogun's ancient capital became a great modern city, 1897–1923*. Harvard University Press.

6

Transitionists: The Dynamics of Transnational Space-Making in Tokyo

The gaijin ghetto and the Pro-Tokyoite are distinct types on the broad spectrum of socio-spatial patterns found in the global city of Tokyo. Whereas these are merely ideotypes on the spectrum, as it has been pointed out already in the previous chapters, such patterns and, thus, the transnational spaces are constantly in change. There is a dynamic element to the transnational spaces or generally to urban spaces, which are especially well represented by the case of so-called transitionists. The transitionists switch their pattern on the socio-spatial spectrum at some point of their migration path to one direction or to another. In fact, these switches, as will be illustrated here in more details, are not solely a story of migration but of life paths at large. This chapter introduces the case of this third empirical type, which illustrate that the socio-spatial diversity must be understood as a process of spatial superdiversification rather than a given state of transnational space. In fact, bringing this spatiotemporal dimension, which has been only recently attracting attention in migration research, this chapter conceptually develops the tempo-spatiality of superdiversification and calls for a more dynamic understanding of transnational spaces in its temporal-spatial contextualization.

S. Yamamura, *Spatial Diversity in the Global City*, Global Diversities,
https://doi.org/10.1007/978-3-031-64725-3_6

There are two major impacts in the lives of transnational professionals which emerged as being decisive for their changes in socio-spatial patterns. One is evidently the career change, which is the overall context of their mobilities, thus, somewhat logical. The other one is the life course change encompassing marriage or at least partnership and family, which is intuitively logical, too. Beyond presenting the narratives and the actual changes in the individuals' attitudes and behaviors, which lead to the socio-spatial patterns, this chapter theoretically adds a crucial temporal dimension to the debate so far on transnational spaces. The conceptual implications thus are twofold: one is the empirical understanding that the temporal and the spatial are intertwined so that we cannot simply talk about the superdiversity but more so about superdiversification as the dynamic process of space-making in the debates on superdiversity. The other implication is the call for analyzing the temporal-spatial context in which transnational migrants, be it as corporate professionals but also other types of migrants (see Chap. 7), are embedded in their global and local mobilities.

6.1 Time Dimension as Structure for Transnational Corporate Mobility Patterns

Especially the transnational corporate migrants moving from one corporate headquarter to another in shorter time spans and high frequency are regarded epitomes of the high mobility caused by the globalization of economy and society. While immobility, which is the other side of the bright coin of global mobility, has been associated with those in society without such economic power, means and resources—those with limited agency on the global stage, what has been missed out in these debates are, as the previous chapters have shown, the actual immobility of these transnational corporate migrants on the local stage.

In fact, this placative image of transnational professionals as the advantaged and privileged of global society and out of scope of immobility debates is a general thread within the debates on the geography of migration. Their privileged position regarding economic power and prominent role in being actors of the global economy is indisputable, yet research has been focusing on this particular aspect, that is, their actions as economic actors and in their business networks. Such privileged mobility discourses tend to be characterized by an underlying implicitness of the transnational professionals' presumed limitless agency in mobility. What has been neglected is their role in the everyday practices as actors of not only the global geography of economy but also as local-level actors in the urban everyday context. Apart from being globally highly mobile individuals, transnational professionals have their specific patterns of social interactions and practices contextualized and localized in space, that is, socio-spatial patterns, showing specific character of mobility within the urban local.

As the transnational professionals' socio-spatial patterns in Tokyo has in fact revealed, the spatial extent of transnational professionals' mobility is double-edged. Although highly mobile globally—especially within the network of global cities—their socio-spatial pattern within such cities is characterized by extreme immobility. The activity space that the transnational professionals cover within the overall 2188km^2 of the Tokyo global city-region constitutes a mere triangular-shaped area of less than 4km^2 in the city center. The previously discussed "Golden Triangle" or the "gaijin ghetto" in Chap. 4 shows basically how much everything from work over family to further leisure activities and social interactions are crunched into. It is a clearly immobile socio-spatial pattern in contrast to their otherwise hypermobile global mobility in their careers: an im-/mobility paradox of these transnational migrants?

As a beginning to this exploration of the time dimension, the classical but still timely approach of analytically distinguishing the temporal contexts according to time geography of Hägerstrand can be useful. It can analytically illustrate how the otherwise globally hypermobile can be indeed constrained not (only) by their preferences but by the structures of corporate context in which they are embedded. The importance of such theoretical take is to reflect not only on the constraints, thus, structures, but also to consider the dynamics between such structure and the agency on the side of transnational professionals. Such structurational approach indeed can then more clearly draw out the dynamics of socio-spatial diversifications.

Reframing this original idea by the founder of time geography into the new mobility context, the three dimensions that time brings into the socio-spatial constellation of the transnational professionals could be transferred and summarized as follows:

Authority Constraint

The authority constraint, or in general context—as it should not be regarded only as a constraining but also enabling factor for the actors—that has temporal-spatial implication for the mobility on a larger scale are constituted by the institutional and power structures of the transnational corporation. The global corporate strategy to deploy their professionals on short- to mid-term assignments according to the processes in the global economy requires high mobility both in distance and frequency, as well as in its limited duration. Comments, such as below exemplify the context in which transnational corporation professionals are embedded in:

> So, financial services is about money, so most people of financial services move to where the money is. So like, why do people rob banks? That's where the money is. You move to where the money is, right. (EM034)

It is not in their direct hands (though they do contribute to the decision-making process in corporate terms) to decide where to and when to move on. It is the flow of money, respectively the economically strategic logics to which they adhere to. Furthermore, it is also part of an increasing corporate strategy for transnational corporations to deploy professionals incrementally to new locations, starting with short-term assignment, placing them in hotels, extending the assignment to a short-term dispatchment, where employees are moved to serviced apartments, and eventually to dispatch them to mid-term migration for two to three years, when family dependents are moved too. This predestined socio-spatial pattern gives an explanation to the corporate structure, respectively the authority context, enabling global high mobility but also constraining the agency locally, thus resulting in the local immobility for the transnational professionals:

> [Referring to the gaijin ghetto location:] Because I used to stay in the Imperial Hotel. And then they moved from the Imperial Hotel into a serviced apartment. And the serviced apartment they put me to ... was in the middle of Azabujuban. So when my wife came over, obviously, there are more Westerners 'round there. (DI014)

With the incremental move to a new location and limited choices of residencies in the beginning, with the time also being limited for further social venturing out during these intensive short-term assignments, transnational professionals are locationally limited into a specific area of the so-called gaijin ghetto, where high-end facilities for global expats are available. Though the agency is provided to venture out of this area when the final migration occurs, the acquaintance with the area and the comfort of the area where similar Westerners or English-speaking residents are concentrated results in many of them remaining in the gaijin ghetto. The corporate strategy of incremental dispatchments, thus, contributes to the high mobility worldwide, yet to the immobility locally.

Another aspect of transnational corporate strategy that constrains the agency of transnational professionals is the increasingly time and cost-efficient relocation of their employees. The relocation, especially enabled by the technical development of the internet and the services provided via internet and emails nowadays, has been increasingly outsourced to intermediary relocation and real estate companies. Whereas transnational corporations used to send their prospective expatriates to their destined city for searching and deciding on their future residences until the 1990s, nowadays, as both corporate members as well as real estate agents point out as a trend particularly enhanced after the global financial crisis, relocations are mainly organized by intermediary companies pre-selecting housings for professionals with emailed information and pictures on which decisions are taking from distance. Such pre-selection as part of time and cost reducing strategy for transnational corporations is a further mechanism of the limited immobility pattern in the respective city as relocation and real estate agencies, as they claim, "know the needs of their global expats" (RE 29).

Moreover, the corporate strategy for this short- to mid-term dispatchment also contributes to the decision-making process regarding everyday practices for transnational professionals. Though given the agency to choose their residence within the city despite the constraints described above, as taking time to explore further places beyond the gaijin ghetto is practically possible, the corporate agenda of the shorter duration of stay impacts the mindset for taking actions:

> My first thought was I assumed that I would live in Japan two years roughly, that was as far ahead as I thought. (DK05)

As this quote from a transnational professional illustrates, the expected shorter duration—which may not necessarily apply to the practice in reality as extension of assignment to further years is not uncommon—clearly changes the mindset for planning the sojourn. It is not the actual duration of the stay that matters, but the prospective duration, respectively the uncertainly of the time duration, on which decisions need to be taken. This leads to pragmatic decisions of ongoing mobility rather than to practices, such as residential and social integration into the local community, which are more adaptable for a highly mobile lifestyle.

Coupling Constraint

The development of internet and communication technologies has unquestionably also changed the everyday practices in business and private, yet face-to-face communication remains a crucial element of semi-formal and informal meetings. Though social media and technologies contribute to increased virtual meetings, transnational professionals still have business trips, especially in the senior managerial level (during the global financial crisis some business trips on lower levels have been reported to be cut down due to cost reasons, but not changing on the higher level), semi-formal meetings with business partners and colleagues are still part of their professional lives, and last but not least informal friendships and socialization mostly develop from such personal interactions. This is where the so-called coupling constraints play a clear role in the shaping of the socio-spatial patterns within the city:

> After-work drinks, we might go to, my office was in Roppongi 1-chome, so we go to some local place there, just for a drink. If we have a team night out ... we wouldn't go too far, so basically because the office was in Roppongi. (ML14)

As this above quotation clarifies, through the locally set "coupling" with colleagues after work, as it is common and not necessarily limited to corporate professionals, the location of such social interactions remains geographically and time-wise close to the company base. As transnational corporations are located in either the central business district or the gaijin ghetto, the coupling of transnational professionals would remain in these respective areas, contributing further to the reproduction of this immobile socio-spatial pattern within the city. Such reproduction of the mobility pattern is also caused by the limited time for leisure and informal activities outside of the corporate context, using mostly after-work setting as the main socialization time. As time for leisure activities are rare especially in the financial industry, transnational professionals tend to couple with those in the similar corporate context, which in turn are located in similar areas, leading to the semi-formal and informal socializations within the same social network. The similar social network is then coupled back again with a relative immobility through the limited occasion for contacts with those outside of the social network and the gaijin ghetto space. Regarding the socialization for private contexts, a professional reflects as follows:

> It's predominantly people that work in the financial industry or at least in an international kind of a company. Rather they work for finance- financial sector. They also meet there. So I would say within those places, you tend to meet the same people. (LD28)

Such coupling pattern within the same social networks then contributes also directly to the choices, for example, of residences, as information are passed on within, limiting consequentially the overall mobility within the urban space:

> So, once I visited a few places and sort of tried to decide between ...(...) I started asking my colleagues at that time like, you know, these are the two places I am looking at, any comments, you know, just in general. (MD13)

Capability Constraint

Last, but not least, the capability constraints overlap with the last argument of the above, that is, the limited time as a resource for the work-life-(im)balance of the transnational corporate life. The physical or practical constraint of having and taking time is a well discussed issue in corporations. In many industries, work-life-balance has taken a crucial role and measures have been implemented to improve the situation, yet it is still a problem within the financial industry. Partly due to the time differences of opening stock markets, working hours are still heavily impacting the everyday lives:

> Normally my working time was very long. Normally from 8 o'clock to 11pm. So that's why weekdays I have no time to contact the other... I don't have private time. (AZ09)

Additionally, the spatial extent of the specific global city, in this case Tokyo, would also be an issue for the capability constraint. Although the subway network is very well developed and the areas well connected, the time needed to access places within the city-region, thus the perceived temporal-spatial constraint often also discussed in mobility research on transportation, is an unnecessary burden for transnational professionals' already limited work-life-balance:

> So I didn't want to live someplace where I had to spend a long time commuting, ehm so that's why I said, okay, all right, I am not doing that. (MD13)

It must be, however, also mentioned in this context that it is a perception of temporal-spatial constraint as the time needed for overcoming the distance between the working place and a prospective residence may not be different in reality than from the chosen residential area. The perceived distance to a location and the perceived time needed is the decisive factor. Areas east of the central business district, which is closer by work by physical distance and commute time, is perceived as being further away and off-the-grid—actually because they are less prestigious working -lass areas and, thus, distant from their lifestyles and corresponding residential areas.

Time as a scarce individual resource for attaining further information and capital to cope with the new environment is also characteristic for the capability constraint embedded in the corporate culture of time-consuming work environment. In contrast to usual migration pattern where migrants are connected to a migrant community and use resources from such social capital, the migration of transnational professionals are primarily embedded only in the corporate structure and social capital outside of these corporate based networks is rather scarce. The issue of language barriers in accessing information, for example, which could normally be eased by social contacts within the migrant network, is common with transnational professionals. Because of the corporate structure on the one hand, with frequent shorter duration moves, but also limited time to attain further resources, such as language skills, the agency to mobility within the respective city can be limited:

> It was an agency that actually does handle a lot of expat type people, but they were pretty good, they spoke English which was pretty important for me… (RE26)

Due to the lack of language skills to access local real estate market, transnational professionals—even if they were free to choose their residence—tend to take the support of real estate agencies specialized in international customers, who in turn, as described above, may be impacting strongly the immobility into the gaijin ghetto where they can achieve the highest profit.

The discussion of the different contexts in which socio-spatial patterns of transnational professionals categorized as the gaijin ghetto type are embedded illustrates how the temporal-spatial approach can disentangle the overall structures influencing these migrants' agency in their mobility pattern. Structural constraints directed by the corporate context are prevalent at different levels: as part of their strategic decisions in the global economy in selecting the destination of the transnational migration, their corporate human resource strategy regarding the dispatchment management outsourcing relocation or the impact on individuals' available resources on time and social capitals for allocating it to non-work-related social interactions. In all these contexts the time component in its various forms appears to be crucial in creating the paradoxical relationship between global high mobility and local immobility.

However, as much as the gaijin ghetto pattern may illustrate the structural constraints to transnational professionals' agency, the case of Pro-Tokyoites demonstrates the agency for mobility, also referred to as motility in literature (Urry, 2016). Despite facing the same structural constraints in terms of authority, coupling and capability constraints of time, as these apply to all transnational professionals operating in a corporate context, agency for mobility can in fact result in a different mobility pattern within the urban space.

As the chapter on Pro-Tokyoites have illustrated (see also quote in the chapter heading), transnational professionals are very well aware of their socio-spatial choices and clearly position themselves from the gaijin ghetto type. Pro-Tokyoites socio-spatial patterns are characterized by the social and spatial overlapping with the gaijin ghetto and more importantly with their venturing out of the Golden Triangle. Geographically, they spread out within the city of Tokyo and diversify their socio-spatial patterns to further locales. It is clearly a matter of agency and active decision. Regarding leisure activities, they proactively seek out local leisure activities, such as non-English-speaking yoga classes or children arts classes in the Japanese neighborhood, Pro-Tokyoites make a different use of the same available time to them than colleagues remaining within the gaijin ghetto, therefore, using their agency to break from the constraints imposed by the structures of the transnational corporate context. However, what are the motivations or contexts behind these different patterns, and how do they actually change over time? The case of transitionists who have changed their patterns can shed more light into this phenomenon in-between gaijin ghetto and Pro-Tokyoites.

6.2 Changes in the Socio-spatial Patterns: Career Change

As transnational migrants who are finding their high mobility as corporate professionals, it is evident that the corporate context is crucial for their socio-spatial patterns. In fact, individual professional circumstances have a strong impact on the socio-spatial pattern and the life course is in particular a factor by which the socialization patterns and, thus, also the spatial behavioral patterns change. In the case of the transnational corporate migrants, many mentioned the change in lifestyle corresponding to their age when they had started becoming more localized, moving out of the gaijin ghetto area:

> Mmh, but once, I guess, we started growing up, we stopped doing that. As I said, you find nice restaurants, cafes, it's all more… much quiet! Much calmer and Aoyama, Ebisu, that's when you start enjoying more, you see, when you hit mid 30s or early 30s. I don't know. (AC007)

> Yeah, exactly, maybe it's age, to be honest, and I didn't have the energy to, I'd much rather be in a bar near to my home with closer friends. And also the whole Roppongi environment is, is, is maybe not as nice as it used to be, not as friendly. (GM012)

Corresponding to the age, however, what these examples do not mention but another one below illustrates well is the promotion in career and higher income, which impacts the actual possibilities of having a specific, usually more cost-intensive, lifestyle.

> *[A]s a student, I lived like a Japanese person, I lived in, you know, a six-mat room in Shinjuku and then when I came back as a professional, I moved back into the expatriate [areas].* (DS039)

This then is also coupled with the desire and need to follow a different lifestyle regarding socialization, such as having friends to visit one's residential home:

> I think the reason was it was a slightly larger apartment. That was a two bedroom whereas the apartment I had in Mita when I first came was a one bedroom. And I wanted a slightly larger apartment. One: I was making more money; two: my housing allowance went up, you know; and three: I wanted to have a place in case friends came to visit. Of course, I was obviously collecting a lot of things that I was buying in different places and I needed some place to put it. (GM012)

Interesting to mention also is the corporate dimension of the socio-spatial pattern. According to one of the transnational professionals, the position within the transnational corporate hierarchy, or at least the nature of the occupation within the transnational corporation, plays a role in the interactions with local people, too. What is described here is that, in his view, the higher executive professionals tend to socialize with each other more compared to those in junior positions, whose tendency to cooperate with local colleagues and, thus, also socialize in leisure time is "not high, but it's measurably higher" than in the case of senior professionals. Hence, this professional attributes the chances of showing a Pro-Tokyoite pattern to the junior position more than higher executives:

> [T]here's, again, there's lots of different communities and, ehm, but like historically, if I look back, largely the senior, the more senior executives, expatriate executives, the way I would define it is, will tend to spend their weekends with each other, across, even across industry. More so than spending any weekend of social time with Japanese colleagues. Right. Whereas the more junior staff that's actually operating and functioning much more like local staff, even though they're born from … you know, raised in another country, are probably more likely…they may … the chances that they will engage socially or do other things outside of work, ehm, with their Japanese colleagues, it's not high, but it's measurably higher. (RL33)

As accounts from other professionals have highlighted, the likelihood that one interacts and socializes with local professionals or builds friendships within the local society is not limited to junior positions. The transnational professionals who have been interviewed and also identified as following Pro-Tokyoite patterns are all higher executives so that the above statement is not applicable. What is important to note though in that quote is that the nature of the occupation, that is, the part on "*actually operating and functioning much more like local staff*" (RL), is the actual decisive factor. With 45 different professionals of transnational corporations within the same financial industry being interviewed, what emerged is the importance of the occupational activity itself. Occupations with higher numbers of encounters and social interactions with Japanese partners or Japanese transnational professionals appeared to result in social interactions also "outside of work," which in turn means a higher likelihood of Pro-Tokyoite patterns.

In professional terms, it was less the position in the corporate hierarchy, but more the nature of the profession itself, such as within the financial industry; being in legal or accountancy consulting of transnational financial corporations in Japan would tend to lead to more contacts with Japanese business partners than a professional in foreign exchange trades with transnational corporations overseas. As banal as it may appear, occupations which required more local social networks and knowledge would tend to require social interactions with local business partners, which again would tend to increase the likelihood that acquaintances and friendships were built outside of the foreign transnationals' networks.

Within the financial industry, consultants of transnational corporations and also other professionals mentioned that the locality and type of socialization outside of office, though still business-wise in semi-formal settings, would be decided according to the needs and desires of the clients. Some mention the need to expand the mobility patterns within Tokyo as meetings were taking place at clients' offices, which in the case of Japanese firms might not be located in the central business district around Marunouchi where the transnational corporation itself is located but further away. Also, occupations which are characterized by "*acting as a liaison with local counsel for local legal issues (...) [where] we do a lot of interacting with local counsel*" (AB_010), or when the professional is "*deeply involved in the Japanese community because 50 percent of my clients are Japanese so, you know, of course, I'm deep into the Japanese society*" (INT_014), involved more occasions where they ventured out from the gaijin ghetto. However, it must be mentioned that accommodating clients' needs cannot be equated with also keeping similar socio-spatial patterns in leisure time. Moreover, there are also cases where more Japanese styles of semi-formal business interactions are outsourced to colleagues with Japanese language or cultural skills. And, furthermore, there are also cases where, similar to the previous case where the Japanese spouse's Japanese friends wished to go out in the gaijin ghetto, the transnational professional consciously took their Japanese clients to the gaijin ghetto for the non-Japanese experience because they were representing a transnational corporation.

What become quite apparent also in other interviews with these transnational professionals are the types of businesses and lifestyles used by them, which are closely connected to the availability of monetary capital. As already described in the section on gaijin ghetto, high-priced luxurious restaurants and buildings are characteristic of these Pro-Tokyoites, too. The difference to the gaijin ghetto pattern, however, is that though in the same price category, the Pro-Tokyoite shows preferences for a stronger local touch. Ambiguous about having the monetary means to meet the expectations of desiring a luxurious lifestyle and socializing with other professionals, the following person recounts the situation as such:

> Yeah, I think my English teacher buddy was very good at being local. I don't think it's…ehm, is it a question of money? I guess, I guess business people have more money to spend and so people think that business people want to go somewhere luxurious and probably they're right. You call me, wanna go to a really nice restaurant? I personally wouldn't appreciate that, but that's me…I want to go somewhere that's local, that's a really good experience, rather than somewhere where they speak to me in English. Ehm, so, if you're asking people to fly in for the weekend, you know, or for a business trip for a week, yeah, I think they'll all stay in Roppongi. I think it's a big mistake. Ehm, all stay in Roppongi, Ginza, somewhere like that. Akasaka, ehm, and so, yeah I think, I think that that's what happens. Go to expensive places. (KW44)

How crucial the socio-economic background is in the socialization and, thus, the socio-spatial pattern is well illustrated with the transnational professional who has moved from Tokyo to Singapore at the time of the interview, but concludes generally on the transnational professional's socialization pattern:

> I think, I, I think the global expat culture tends to socialize amongst itself. That's not to say that they don't have contacts in the country that they are in, because, you know, I think anyone that enjoys being overseas, if they're not getting involved in the local culture, then, you know, what's the point of being there? And, quite honestly, what they tend to do is, or we tend to do, is you begin socializing with people of a similar social level, if you will, people of a similar corporate level, who are also locals to the particular place where you live. So, I wouldn't say that there is a, ehm, invisible barrier between the expats and the locals. I think living anywhere, whether it was Japan, which I think was one of the greatest places I've ever lived, or Singapore, you know, it's enjoyable to get to know the local people. (GM24)

What this quote shows is not only the closely knit socializations among the foreign transnational professionals (here: "expats") but also those with the local professionals, whereas the "invisible barrier" between any social groups is perceptively not to be drawn according to being foreign or local, but rather along the socio-economic levels of the professionals.

As a matter of fact, how much the socio-spatial patterns rely on economic capital becomes clear in the case of those who lose their socio-economic level through either loss of their position or even job. As real estate agents described bluntly: "*the more money the more central" and "the worse the salary, the farther away from the center*" (Real Estate Agent 02). Such spatial dynamics is closely connected to social networks and practices. The interesting finding regarding such socio-spatial dynamics caused by financial capital is the multidirectional change. According to another real estate agent specialized in the same customer base, it is not only the move from the center to more "peripheral" areas (NB: what is perceived peripheral is still in within the twenty-three wards of Tokyo, thus, central in common understanding). There is also a vertical movement when these tenants lose capital. Retrospectively, the agent remembered that several were not able to pay their high rental in the lighthouse high-rise buildings, such as Roppongi or Atago Hills, but still earned enough for these upper-class housings and to avoid changes in the address, they moved down the floors within the same building. The multidirectionality of socio-spatial mobility in accordance with socio-economic downward mobility contributes to the complexity of these socio-spatial phenomena.

6.3 Changes in the Socio-spatial Patterns: Life Course Change

Apart from these professional career and corporate contexts in which transnational professionals are individually embedded and by which their socio-spatial pattern alternates, the other important impact is the life course changes in their private lives, as it is the case with those with gaijin ghetto patterns as well as the Pro-Tokyoite.

One of the interviewees mentioned the different lifestyles and patterns according to these private circumstances as follows:

> So, a lot of people live very different lives here, particularly if you're … you're married with children, if you're married with no children or you're single. Those three groups live very, very different lives here. I mean, if you think about the business community, which can be quite small here, a lot of the children are going to the international schools together, playing in the same sports teams. I think there is … that group is pretty closely knit or connected. And I think there is a … there is a group where I've seen quite a few … you know, maybe younger international folks here that are married who, maybe are not executives but they're here kind of mid-level roles, and they live very different lives. (GH40)

Marriage/Spouses

Regarding the marriage or being married as a life course element, there are contrasting differences to the ideal-typical case of the gaijin ghetto pattern, where the marriage had resulted in an enhancement of the social

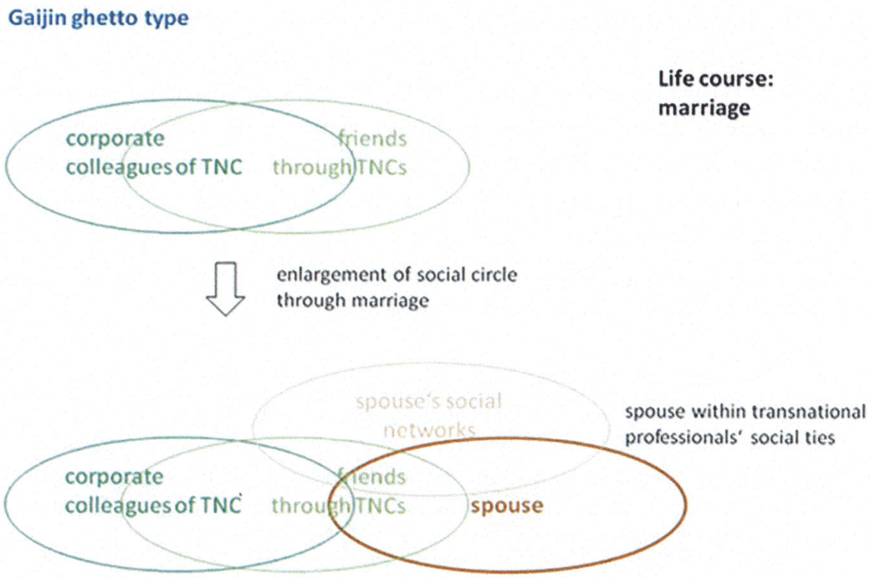

Fig. 6.1 Gaijin ghetto—life course change of marriage (author's own)

networks with overlapping of the social circles (Fig. 6.1). In abstract terms, the lifestyle already changes drastically with age and the social circumstances. As one of the interviewees, living in Singapore at the time of the interview, explained about the change from serviced apartments or even longer sojourns in hotels:

> [W]e settled down. I've got family now. Actually, my wife is Japanese. And I've got an eleven-year-old daughter. But, yeah, I can't be so transient to live at a hotel. So, we've got an apartment here in Singapore. (GM012)

When a transnational professional marries or is married, the spouse enlarges the social circle of the professional who is—due to not uncommonly limited leisure time through work in the financial industry—often constrained to social interactions with corporate colleagues, from which circle friendships also emerge. The spouse adds more social contacts in general, as in any other case of partnerships, yet in the case where the spouse is closer to the professional's own social network, the circle remains relatively dense, that is, still closely knit in the gaijin ghetto circle.

What is typical in the case of these Pro-Tokyoites regarding the social environment is an enlargement of the social networks and, thereby, the enlargement of their activity space (Fig. 6.2). These transnational professionals tend to have spouses who enlarge their social circles, be it through the spouse's local ties by marrying Japanese, having a Japanese-speaking spouse, or by being married to a person who has a stronger interest in localizing despite a lack of language skills.

> I mean the other aspect to it is my wife … my wife is also working here. And she is working with an English language organization, she is a psychologist. And so she is working three days a week as it is. And it is kind of funny because we … so […] yeah, she works in an English language organization but she has a lot of … the majority are Japanese clients that happen to speak English. (AB010)

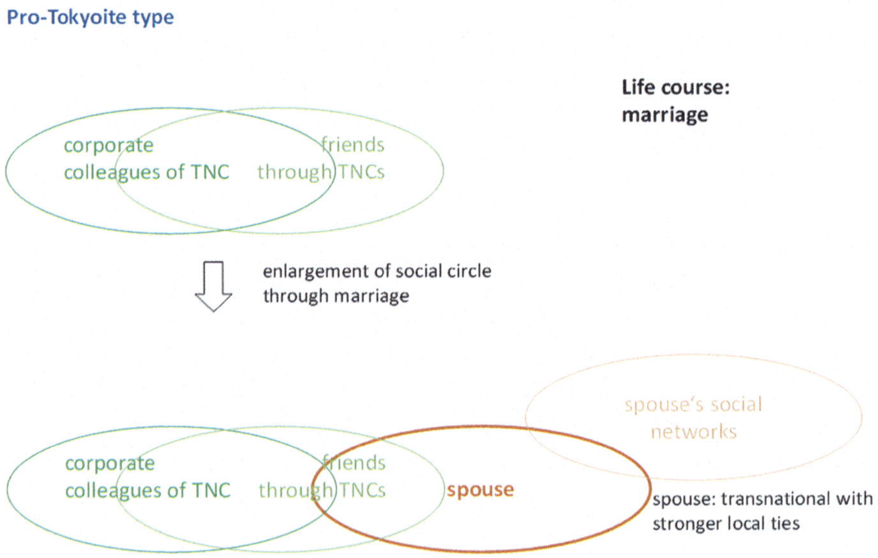

Fig. 6.2 Pro-Tokyoite—life course change of marriage (author's own)

Even with a US spouse, by the circumstance that her colleagues are mostly Japanese yet with English skills, the social circle is extending to Japanese people and by recommendations and invitations to more Japanese areas, the transnational space is also extending to larger areas. Schematically, the enlargement of social networks is illustrated below. Rather than overlapping with already existent social circles, the spouse contributes to an enlargement of the circle. What were fairly typical in these cases were the independent activities of the spouses, generally the wives,[1] apart from their roles as school mothers. Usually, those who extended the social circles were either professionally active as in the previous case where the wife, a psychologist, worked for an international organization or already had some social ties in Japan through their Japanese backgrounds or previous stays, whereas their activities and attitudes also depended on the type of occupation. For example, a spouse who was independently working as an English teacher in an international school would show little Pro-Tokyoite patterns compared to one who was active in adult learning with Japanese students.

Children/Family

The other life course impact, apart from the spouse, is the extension of the family through children. The interesting part of this is that some transnational professionals perceive having children to be a reason for leaving the more Japanese style of living as a Pro-Tokyoite, primarily for convenience reasons as they say:

> The first time living here as a professional, before that as a student...I didn't want to live, yeah, I wanted, it was the first time my wife was living in Japan, so I wanted something that was a little more of a Japanese experience (...) we didn't have children then, so that's the kind of experience I wanted and then when we came back, we were in New York for two years, and [when] we came back again in '94, my first daughter was born right

[1] It can be generalized with wives as only two of the forty-five interviewees were female, of which one was single and the other one married with a husband who also worked in a transnational environment.

> after we came, (…) I didn't really feel the need to live, you know, very Japan kind of [life]. (DS16)

At the same time, if the pattern of a gaijin ghetto type continues with the extension of the family, meaning the addition of social circles through children between the professional and the partner, the same social circles can be enhanced rather than expanded (Fig. 6.3). This would be the case with those gaijin ghetto professionals who choose to socialize in the clubs, such as the TAC, or spend their leisure time in corporate clubs or doing activities offered by the Consulate. By choosing to socialize in such gaijin ghetto locations and also concentrating on the social circles of the international school, the socio-spatial pattern results in an overlapping of social circles. Parents of school children become befriended, who in turn are often also involved as corporate professionals of transnational corporations; the same also applies with the leisure activities where friendships are built more in the enclave rather than outside of the gaijin ghetto.

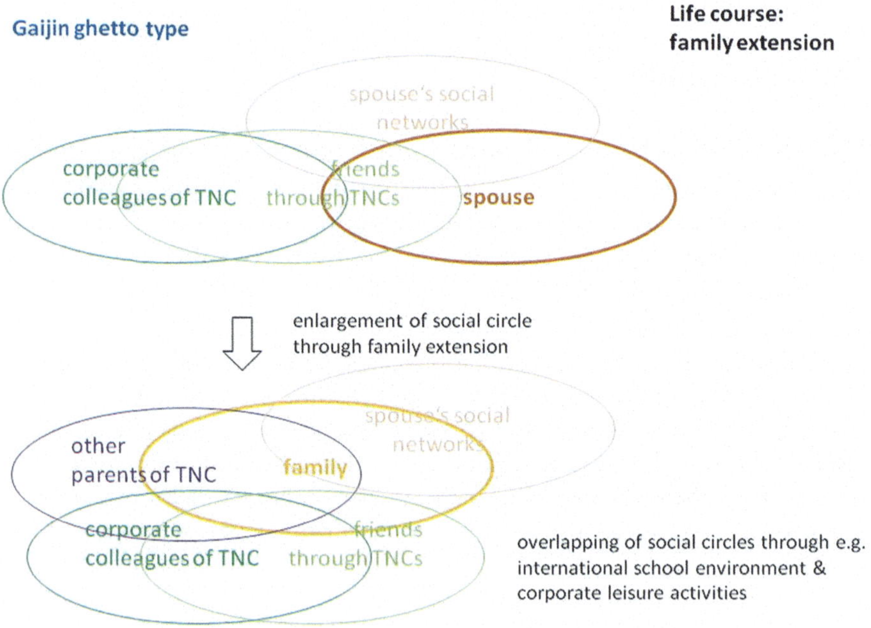

Fig. 6.3 Gaijin Ghetto—life course change of family extension (author's own)

As one of the transnational professionals who have moved from Tokyo to Singapore by the time of the interview explains, this tendency is strongly connected with socio-economic status and the lifestyle chosen. Though taking Singapore as an example, the same pattern applies to the behavior in Tokyo, too:

> You know, I think there clearly is something to say about the socioeconomic level of your friends and acquaintances, because clearly, you know, if you make decent money in investment banking, there is no reason that you're not going to talk to somebody who is in a different industry that enjoys the same things you do, that goes to the same places you do, whose children probably go to the same schools as your children, so I don't think there is, you know, any kind of wall around people socializing from a corporate basis. I think it's more, if you see people doing the same things, going to the same places, having similar experiences, you tend to gravitate towards them.
>
> (…) I mean, clearly, having, having things in common, you're going to run into people here at the restaurants as you go out at night. You're going to go on holidays and invariably you're going to be going to the same destinations. You know, Bali, Phuket, Bangkok, Mumbai, wherever it might be, you'll see them on the airplane. They tend to, especially in Singapore, to take the same airline, which is SQ. They'll all see each other in the same airplane, association lounges, their children will go to the same schools, so the mothers will meet the other mothers of these kids at those schools. So there is a lot of ways that you continually interact on a daily basis both within Singapore and outside of Singapore. (PL13)

The main aspect of having a family extension through children is the expansion of the social networks through the children's involvements in the international schools. It must be noted that none of the foreign transnational professionals chose to place their children into the Japanese school system because of its incompatibility with further schoolings in their future career paths.[2] Through the socialization of their children as

[2] There were only a few exceptions: in one case, a European national school was chosen over international schools and also in one binational family where the professional had the long-term prospect of staying in Japan a local Japanese school was chosen; exempt are Japanese families with returnee transnational professionals.

well as themselves as parents in the school community, their social circles are, similar to the gaijin ghetto parents, expanded. However, compared to the expansion that the Pro-Tokyoite experiences through a spouse with stronger local ties, the expansions of social circles are limited through the international school environment. Some professionals even point out that it is through the school that they have become more involved in the expats community because the children are involved in the international schools where parents of similar corporate backgrounds can be found again. Schematically, the effect of family extension, thus, would look as follows (Fig. 6.4):

Also, regarding the integration of transnational families with local (Japanese) friends, the choice of school in addition to keeping a specific lifestyle as a transnational professional after the family extension appears to play a role in their increasing detachment from the local society, too. Though originally showing stronger Pro-Tokyoite patterns as a couple with stronger ties to the local social networks through the Japanese wife's

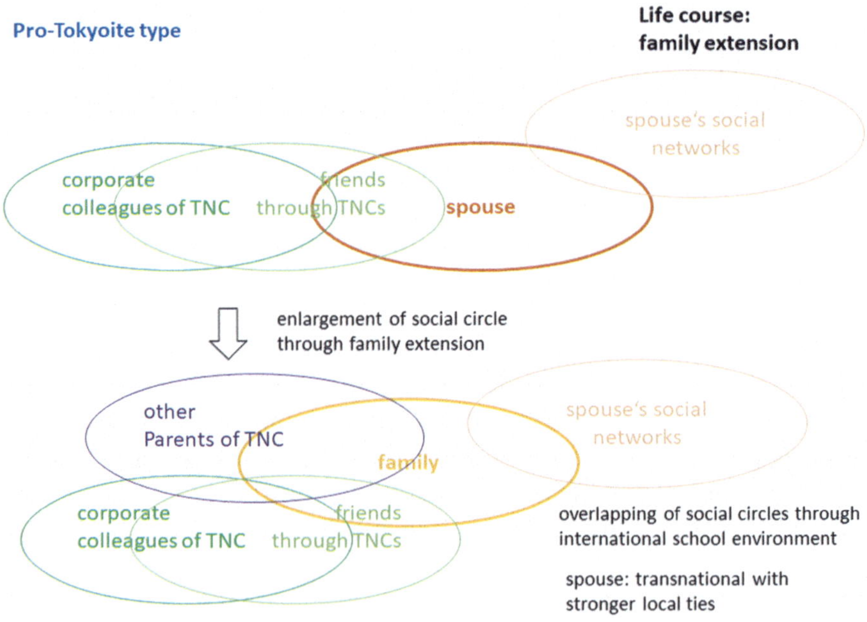

Fig. 6.4 Pro-Tokyoite—life course change of family extension (A) (author's own)

friends, the transnational professional recounts the change after family extension:

> I think the gap between our lifestyle—downtown Tokyo, big apartment, kids in international school—and the lifestyle that her friends from college moved into when they married suburb men and lived out in the suburbs, has just grown to the point where it's not easy for them to keep in touch. They can't get together and have coffee, because they live an hour and a half apart. (BL28)

In fact, the choice of school has a strong impact on the transnational professionals' socio-spatial patterns and that of their families. With national schools also present in global cities, what has been reported on them is that also at the parents' level, a more nationally oriented socio-spatial pattern may occur through the school involvements.

> Well, there are also many, many Europeans. But I don't interact with them as much. It's interesting. [Why?] I don't know. Maybe language perhaps. There are lots of, you know, French and Germans—and what other groups have I seen? French and German are the main ones. Through the American Club. And of course they speak English, too. But just haven't tended to mix. It's interesting, I don't know why. Maybe again because of where their kids are. The kids of French expats tend to be in a French private school, and so we just don't meet except to say hi in the halls of the American club, maybe. Kids have a lot to do with how expats meet, I think. (BW18)

Having said that all children of transnational professionals are normally registered in international schools, there are—apart from those who go to a local school—those transnational professionals of Pro-Tokyoite pattern who would consciously choose the school according to the connection to the local culture, pointing out the reason for the choice of that particular international school as follows:

> And it has been [...] a continuing dialogue for us with the school that she is in. Which of the international schools that we looked at tends to be the

> most sort of Japanese of the international schools. The Japanese culture is a big part of the curriculum. The Japanese language is a big part of the curriculum and the population of the school has changed a lot in the last three or four years. (AV10)

Furthermore, whereas gaijin ghetto pattern parents mention the international schools as one of the main reasons to be limited to socializing with the same social circles and several Pro-Tokyoite professionals mentioned being more limited to the international school community, the difference of Pro-Tokyoite parents who maintain their pattern would be that they seek to stimulate their children differently, despite registering them in the international schools. The following example illustrates how the transnational professional consciously suggests his child takes an extracurricular arts class in a local arts school:

> Actually, it is funny we are talking to our daughter, our older one, about joining an art class that is run through a Japanese school and Japanese part of town. And I think she will be doing that and it is part of our goal for her to get more exposure with Japanese language and culture. (LK25)

In such cases, as well as in cases where children go to a local school, transnational professionals actually mention the children being the reason that their social circles are broadened and sometimes even go so far as stating that they experience the local culture only through them, which stands in contrast to those Pro-Tokyoites who somewhat return to their comfort zones for their children.

> Well, having said so, I worked and my husband worked, although it wasn't long hours, it's very difficult to interact with the local community, so it was really more through my son's school, kindergarten or kid's friends' parents and so forth that we had any connection to the community. I think it wasn't as much as you would expect. (JT09)

Schematically (see Fig. 6.5), in these cases, where the Pro-Tokyoite transnational professional maintains the pattern despite or because of their children, respectively, their way of raising their children, the social circles

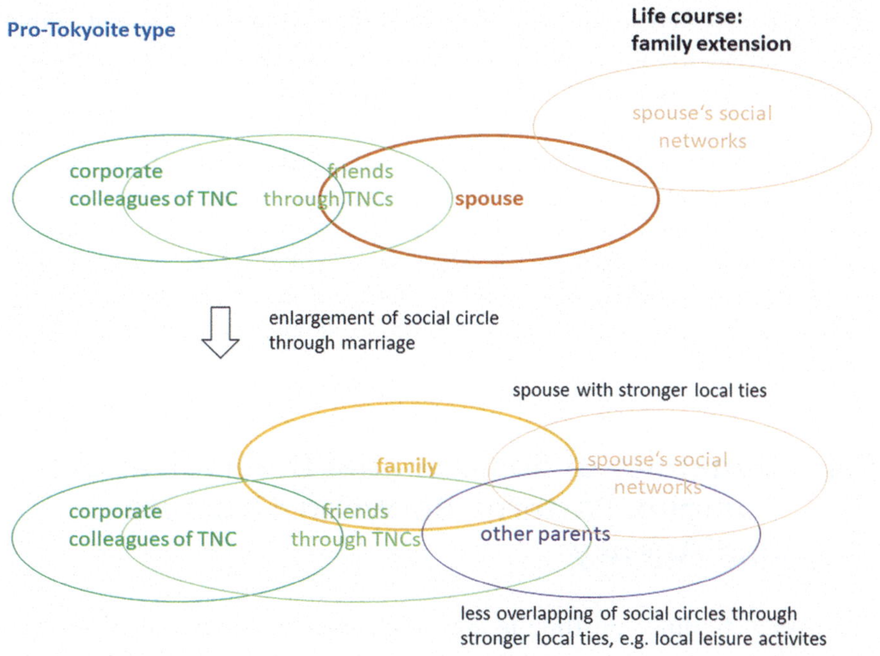

Fig. 6.5 Pro-Tokyoite—life course change of family extension (B) (author's own)

consequently expand as the overlapping of social circles with gaijin ghetto parents are reduced either through the different type of school (local or more Japanese international school) or at least when they involve their children in local leisure activities instead of the corporate or consulate organized clubs.

Second-Generation Life Course

In addition to the life course aspect of having children for the transnational professional's socio-spatial pattern, the life course change of their children has to also be considered as a factor influencing the parents. With growing age, teenagers would start uncoupling from their parents, schools would reduce social events for middle school or even high school children, and consequently reduce social interactions among school parents, too. Asked about the difference if the children were still at home or even about the difference the age of the children made to everyday practice, one transnational professional responded:

> A huge difference. And it made a bigger difference when they were younger. When my daughter was age three to eight, or so, so early, early school, there was a lot more parent involvement in stuff at school. And that's when we met people and when my wife especially met all the other mums. And now that my daughter is in middle school and my son is almost nine, there's a lot less of that going on, so we're not meeting people as much. (LO28)

6.4 Dynamics of Socio-spatial Diversification: Complex Patterns Between Structure and Agency

The overall findings on the life course changes in transnational professional's lives and their socio-spatial patterns, but especially the last example, show the complexity of the contextualization of these patterns. By different individual circumstances of the transnational professional, as well as those of the spouse who comes into the professional's family life, and in addition the abundance of choices that are made for the children and their changes in behavior during their life course, a myriad of different socio-spatial patterns results for transnational professionals. As can be seen, the schematic illustrations are ideal-typical and cannot be otherwise visualized due to the diversity. It must be kept in mind that gaijin ghetto and Pro-Tokyoites remain the extremes on a large spectrum of socio-spatial patterns and should not be understood as a rigid and fixed category of people, let alone community. Despite these diversities of patterns, it still remains interesting that there are these main patterns in socio-spatial characteristics within the transnational professionals.

The complex overlapping and expanding of social circles through spouses and children in one's life course is also well illustrated by the following quotation from a long-term transnational professional who has a Japanese spouse and children at an international school. In addition to the social circles spreading through family members, he also touches upon the difficulty of maintaining his social contacts over time, whereas the time dimension refers to his immobility in contrast to other short- to mid-term expatriated professionals:

> It's interesting. It's ... there are definitely two groups. The biggest group has always been foreign expats. Mostly Americans, some Canadians, who have been sent here on 2–5 year joints. And we usually have met through the kids. So, their kids are in the same class with my kids and the parents meet at school events. And we've had a big circle of friends that came from that arrangement. And then one of the really hard parts about being an expat is they always leave. I have been here forever and I keep meeting new people that after three to five years leave, and they leave and they leave. And I'm finding I'm running out of emotional energy to try and make new friends, knowing that they are going to leave again in two or three years. So I have this wonderful Facebook community spread around the world of people I know, but increasingly few friends left in Tokyo and, as I said, less energy to go out and start again. So that's, that's a problem. The other group of friends is Japanese, but almost exclusively through my wife. My wife and I are friends with her friends and their families. (IN02)

These presented dynamics of transitionists emphasize that the spatial diversity in cities is characterized by their processual quality. Transnational spaces should not be perceived as a finished or at any time stationary space, but as a dynamically changing, yet locationally tangible space. The spatial diversity in global cities caused and lived an increasing number of transnational migrants can be best described as a spatial *superdiversification* in a processual sense. The empirical research has also shown that, in this context, there is no one factor within the diversity of individual transmigrants which causes or explains the transitioning from one socio-spatial pattern to another. This may appear to be a rather disappointing finding for some at first sight, but reflects adequately the complexity and non-mono-causality of societal dynamics.

Moreover, returning to the theoretical question regarding the apparent im/mobility paradox, there are even further dimensions in this complexity of superdiversification and mobility phenomena. Whereas the corporate context as the structuring element of transnational professionals' lives, impacting the shaping of their socio-spatial patterns in their global mobility appears evident, the mechanism regarding their agency and their actions in the local socio-spatial behavior is still unexplored. The key to understanding this context appears to lie in what Giddens already pointed out as "the 'situatedness' of interaction in time and space" (Giddens, 1984: 110). Though structuration as a lens to analyze set of constraining and enabling elements of mobility has been explored extensively, the issue of the temporal in this context has been not yet thoroughly explored. Whereas Giddens had already identified time geography as a crucial component of society, and time geography by Hägerstrand (1970) with its origins in the observation of migration further developed into debates on the time and structuration led by Pred, the role of time constraints in the socio-spatial mobility pattern is underexplored. The temporal-spatial approach to mobility patterns though can disentangle the complex interplay between the corporate structures and the agencies of transnational professionals which lead to the apparent im-/mobility paradox.

In other words, in terms of time geography, the difference between the two patterns including also transitionists is that Pro-Tokyoites make use of their agency to break from the constraining nature of the different time constructs, demonstrating that with some proactive effort (otherwise conceptualized as motility), it is possible to overcome the constraints provided by their corporate and social context. However, pro-Tokyoites still remain somehow limited in the patterns, too (but with slightly "larger" boundaries). This suggests an interplay of structures and agency in understanding these patterns accompanied by a strong role of time dimensions in the interplay.

As displayed schematically in Fig. 6.6, this means that the overall corporate structure in which transnational professionals are embedded is tuned by the three different time dimensions of authority, coupling and capability constraints, resulting in mobility patterns with different spatial extent. The corporate context especially explains the high mobility on the global scale of all transnational professionals. The local immobility is embedded in the same constraining structures, too, but with the agency for mobility (motility), different socio-spatial patterns (yet still on the same high global mobility level) can be observed breaking partly out of

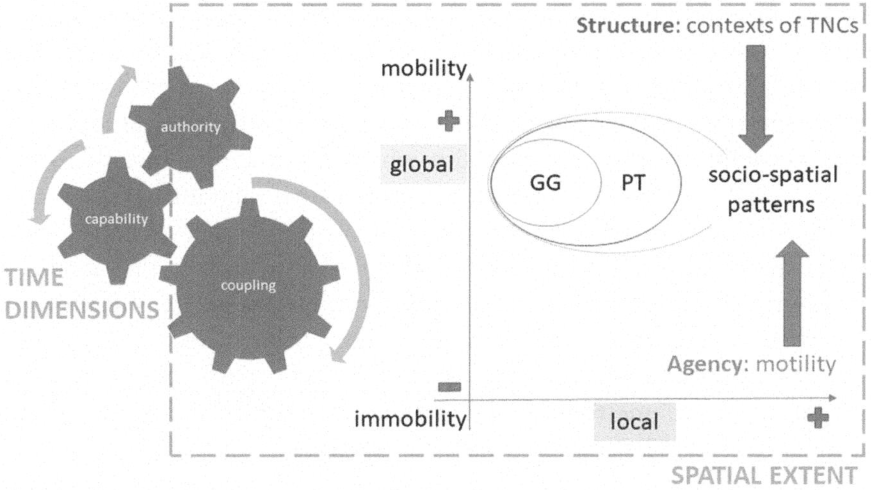

Fig. 6.6 Time dimension in the structuration of mobility patterns of transnational professionals (*TNC* Transnational corporations, *GG* Gaijin Ghetto, *PT* Pro-Tokyoite) (author's own)

the smallest gaijin ghetto (GT) to Pro-Tokyoite (PT) spaces. The spatial extent of local mobility thus is dependent on the individual decisions of transnational professionals to overcome these structural constraints given by the corporate context. This evidence of agency is in the action, which, in this case, is the different socio-spatial patterns observable between gaijin ghetto and Pro-Tokyoite patterns.

The integration of time geographical concept into the overall dynamics of structuration helps uncover these interplays of structures and agency in understanding socio-spatial patterns of mobility and their apparent paradox for transnational professionals. Whereas structuration approach helps explaining the overall dynamics between corporate structures and agencies of transnational professionals, as not only economic but social actors on a general basis, the time dimension brings a more subtle and nuanced explanation to the differences between the existent socio-spatial patterns, in particular on the local socio-spatial extent.

As there is no one single explanatory factor for the differences in socio-spatial patterns, the different usage of agency in overcoming particularly the strong temporal constraints of the structure appears to be the key to understanding this phenomenon. The time geographical perspective is a crucial aspect of the transnational corporate context less dominant in other migration groups. Disentangling these corporate structures by analyzing the time dimensions thus helps explain the paradoxical aspect of global mobility and local immobility, which appears to be specific to the socio-spatial patterns of transnational professional migrants.

From these presented empirical results on the local level of transnational spaces within Tokyo, the next chapters now venture into a reflection of these socio-spatial dynamics by zooming out geographically and also with an excursus from transnationalism from below, giving more facets of transnational migrations and their socio-spatial phenomena in the larger Tokyo metropolitan region, and bringing them to a more abstract level of debating the transnational space.

References

Giddens, A. (1984). The construction of society. *Polity*.

Hägerstrand, T. (1970). What about people in regional science? *Papers in Regional Science, 24*(1), 6–21.

Urry, J. (2016). Moving on the mobility turn. In *Tracing mobilities* (pp. 13–23). Routledge.

7

Transnational Space over the Global City-Region

The one distinct characteristic of the socio-spatial patterns of transnational professional migrants, and by that their transnational space, is their transnational nature, that is, the spatial expansion of their patterns beyond national borders. Before discussing the spanning of the specific transnational space beyond such national borders on a global level and embedding it into the broader context of global city networks, there is also the meso-level between the local urban and the global dimension of the transnational space which needs to be discussed. As could be seen in the previous chapters, the transnational professionals' social practices are not strictly limited to the inner core city center of Tokyo, but also show dynamics beyond the city limits, namely, the global city-region.

The global city-region signifies the spatial phenomenon of "distinctive sub-national (i.e. *regional*) social formations whose local character and dynamics are undergoing major transformations due to the impacts of globalization" (Scott et al., 2011: 1; emphasis in original) whereas in contrast to the "older regionalism" these regions are thought to become increasingly independent from national states but more so subject to the occurrences of the global stage. The global city-region as an analytical unit has been dedicated special interest among scholars of global city

S. Yamamura, *Spatial Diversity in the Global City*, Global Diversities,
https://doi.org/10.1007/978-3-031-64725-3_7

research (cf. in particular *Global City-Regions. Trends, Theory, Policy*, edited by Scott, 2011), and undeniably also has relevance for the spatial diversification of the global city Tokyo. It is these dynamics and social (-spatial) transformations of the global city-region which are also reflected in the transnational space-making beyond not only national but also urban administrative borders.

However, before delving into the empirical evidence of the socio-spatial diversification in the metropolitan region of Tokyo, a brief recapitulation of the definition of this particular "region" is required as this is in fact already a somewhat controversial, or at least rather vague, issue within global city-region research. Despite different understanding and interesting disputes on the definition, one aspect which all definitional approaches of the scholars have in common is that the "region" and its borders are defined by its functional context rather than administrative. Whereas Scott refers to the social formations and transformations on a sub-national scale (2011), the core idea of this "territorial scale and an analytical construct" (Sassen, 2011: 78) derives from economic geographical and regional development contexts of the global city, in which industries and firms locate in the suburbs or even broader regions surrounding the global city. In this context, not only the distribution and functioning of economic actors are being discussed, but also matters of governance and regional policies since these global city-regions are characterized by their attachment and direction to the economic impulses caused by globalization (see also Scott, 2011).

Sassen discusses the contextual difference of the global city versus the global city-region. As previously laid out in her work (Sassen, 1991), the argument toward a strong social polarization and a diminishing of the middle class from the center is dominant in the global cities' context (which in the meantime also heavily contested[1]). In contrast to that, Sassen later suggests that the global city-region as a concept with its nature in accommodating the social middle class and economic sectors (firms) could be used to analyze "a more even distribution of economic benefits under globalization" (Sassen, 2011: 78). The scholars of global city-regions, thus, do not only refer to the purely economic dimension of

[1] See e.g. Fainstein (2011), White (1998), Hamnett (1996).

regional development but also attempt to integrate societal aspects into the concept, too.

Yet, although the definition proposed by Scott (2011) refers to social formations and transformations, the main research field and perspectives of the research on global city-regions remain primarily economic geographical and do not thoroughly discuss the socio-spatial or socio-cultural aspects of spatial diversification in these regions. The social geography is being discussed as one of the agenda of the global city-region debate, touching upon issues of migration and the "increased cultural and demographic heterogeneity" induced by it (Scott et al., 2011: 18), but empirical material on the how and what exactly is occurring cannot be found. As a matter of fact, empirical backup to these debates is mostly quantitative economic and social data on cities, regions and city-regions, but they do not break it down further into the sub-regional level (partly due to limited availability of data), and when such cases are mentioned, it remains rather an eclectic patchwork of examples taken from case studies, for example, in the case of the "definition," respectively, description of the polycentric city-region elements identified by Hall (2011: 73-74).[2] This is in large part due to the conceptual framework of not defining the global city-region by its administrative borders but by its function, in particular by its "internal linkages" (Hall, 2011: 72), which in turn becomes empirically problematic as the statistical data on which these debates are carried out are only delivered on administrative units. Therefore, this apparent contradiction in the empirical evidence, especially as an inter-regional comparative analysis is intended, and its underlying concept is met with the approximation for the unavailable data.

There is in fact an inherent difficulty in discussing the global city-region of Tokyo because of this definitional problem. The metropolitan

[2] While describing the geographical structure of global city-regions, Hall lists six types of sub-regional areas, bringing samples of diverse districts in major global cities. The selection is not necessarily exhaustive, for example, for Tokyo only four of the six types are identified (including two of them being misspelled) and the empirical value in some cases is also questionable, for example, Shin-Yokohama would historically not be considered "'outermost' edge city complexes" to the Tokyo global city-region but the outcome of a strategy of establishing the 'second city center' (City of Yokohama 1989), also later designated to be one of the twin core with the waterfront bay area of Yokohama (City of Yokohama 1994), to the megacity Yokohama; the planned Tsukuba Science City located approx. 50 km northeast of Tokyo would be the more adequate R&D center to Tokyo.

region of Tokyo as an administrative unit already shows a diversity of definitions and data bases. In the case of Tokyo, it is not only a matter of polycentrism of the Tokyo Metropolis (the metropolitan prefecture) itself with its 23 cities or special wards (subsumed as the central City of Tokyo) but also the further 39 municipalities on the outskirts. It is also problematic as the rougher National Capital Region encompasses a further six adjacent prefectures including the metropolis Yokohama, further industrial cities around and in-between Tokyo and Yokohama, but also rural and sparsely populated areas. To take into account the labor population attached directly to Tokyo city center, it is possible to look at the Tokyo Metropolitan Employment Area which statistics base on the commuters to the twenty-three special wards. An alternative for incorporating the multiple cities within the metropolitan area, so the global city-region, would be to consider the Kanto Major Metropolitan Area, which encompasses all municipalities commuting to one of the five so-called designated cities (Yokohama, Kawasaki, Sagamihara, China and Saitama) or the twenty-three special wards. However, the commuting population does not cover the locational issues of the regional economy, that is, the location of firms and industries. The methodological question arises about the selection of the administrative units for quantitative data and its conceptual implications for the analysis of the respective global city-regions. In any case, a simple juxtaposition of the sub-regional elements of the global city-regions as suggested by Hall (2011: 73–74) which are based on urban districts within Tokyo and not sub-regional units is more than problematic.

Apart from the problematic methodological approach to the concept of the global city-region, there is also a lack in the empirical base on the assumptions, in particular regarding the social aspects. In the edited book by Scott, which attempts to bring together not only a variety of the most influential scholars in the debate on global cities but also a variety of topics by which the global city-region phenomenon is supposed to be evidenced, Fainstein discusses her observations on social inequality from her previous research (Fainstein, 2001), questioning the global city hypothesis of the "disappearing middle." The linkage between the global city-region and social inequality is seen in the income distribution—with the tendency for an accumulation of high-income residents in the global city-region as a consequence of the rising living costs and the movement of lower income and non-global firms to the periphery—and also, due to the reliance and focus on the financial economy and its markets, an instability of the residents in such global city-regions (Fainstein, 2011). Yet, similar to the other authors, the data base on the administrative cities and aspects of migrants remains not fully described. Specifically, the "empirical evidence" on the Tokyo global city-region is limited to secondary data taken from Sonobe and Machimura (1997), who base their discussion primarily on the occurrence of the bubble economy in Japan. The context of foreign workers, that is, (unskilled and lower-skilled) labor migrants, as being concentrated in the industrial areas and, thus, not in the core urban centers is not to be contested as the service sector for which transnational professionals migrate to the global cities is located mainly in the city centers, but findings on the extent of the "increased cultural and demographic heterogeneity" (Scott et al., 2011: 18) by migrants in these global city-regions cannot be easily deduced.[3] However, as the excursus on the mechanism of global-city-(region) shaping of the migration industry from below and above will show at the end of this chapter, the socio-spatial diversification from that front appears well justified.

Though the observation is certainly correct with labor migrants being generally spatially more concentrated on the outskirts of the city center[4] and close to industrial areas compared to transnational professionals centering in the global city itself, the dynamics of socio-spatial contexts of transnational professional migrants in the global city-regions have been left totally out of sight. When it comes to transnational professionals, the focus has been limited to economic spaces and social connections within economic activities inside the borders of global cities. However,

[3] Waldinger is the other author in the same Part VI titled "Social Inequalities and Immigrant Niches in global city-Regions" (edited by Scott, 2011), but the chapter primarily discusses the concepts of ethnic enclaves and immigrant niches and neglects the contextual embeddedness of this phenomenon in the global city-region. The global city-region is touched upon in the introductory paragraphs, but apart from generic use of the term global city-region instead of metropolises, it is not further conceptually discussed and the empirical data refer to the large cities rather than global cities as done so in his previous works.

[4] With reference to literature on urban poor and areal deprivation, for example, urban slums and ghettos, such a general statement is hardly tenable but, even so, in relative terms to the transnational corporate professionals this can be generalized. At the same time, as discussed in Chapter II, historical migrants of Chinese and Korean backgrounds are very well represented in the inner cities.

transnational professionals are not only economic actors as part of transnational corporations, but also actors of urban and regional economy in the broader sense as transnational migrants, involved in the housing market and everyday social practices embedded in the urban and regional contexts. The question thus lies in the neglected aspect of the nexus between the global and the global city-region spaces via the agency of transnational financial professionals: What then are the socio-spatial practices of transnational professionals and their connections to the dynamics of the global city-regions?

Apart from this chapter being a crucial step of zooming out from the intra-urban spatial diversification dynamics within the global city of Tokyo to a larger context, it also contributes to the global city-region debates in three aspects: (1) by bringing empirical evidence not from the methodologically and conceptually problematic quantitative data but from qualitative research, it contributes to further discussions on empirical grounds in the case of Tokyo; (2) on the basis of these empirical findings, it also brings some contours into the "patterns of social stratification, intra-metropolitan income distribution and demographics, and ways of daily life" which are substantially marked by the "forces shaping the emergence of global city-regions" (Scott et al., 2011: 18), as depicted by the global city-region scholars; yet, (3) focusing especially on the latter "ways of daily life," that is, social-spatial mobility and practices (partly also socio-economic mobility) of those who have been assumed as being restricted to the urban core, namely, the transnational professional migrants. The three trends within the researched transnational professionals, which will be further elaborated below, provide evidence to the production of transnational space which is not only crossing national borders but also the city borders of the global city Tokyo and, hence, evidencing the clear relevance of this group of transnational migrants for the dynamics within the global city-region (Fig. 7.1).

The extents to which transnational professionals contribute to the dynamics in the global city-regions beyond the city borders of Tokyo are manifold, yet can be distinguished according to the three patterns discussed in the previous chapters. All three patterns, that is, the gaijin ghetto and the pro-Tokyoite as well as the transitionists, all venture out of the limited urban space of the global city. However, as can be deducted

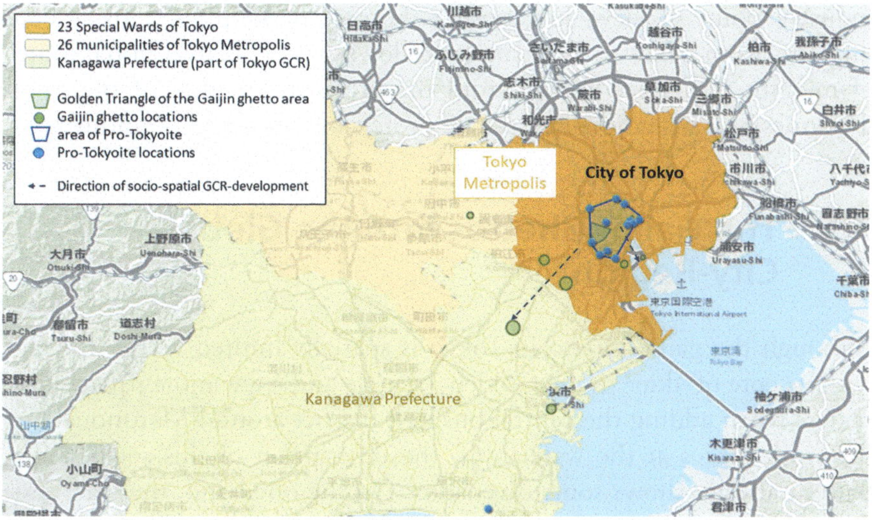

Fig. 7.1: Global city region of Tokyo (Yamamura, 2019: 516/CC BY 4.0)

from the socio-spatial diversities described further previously, the spatial extent in its geographical sense of both the area and the locations, and also the social contexts and the larger structural embeddedness, is completely different. Though all three patterns will be discussed, in the case of the transitionists it must be noted that these are—in contrast to those showing mobility on the spectrum of the transnational socio-spatial patterns—moving not only socio-spatially but also socio-economically.

In the following part, the previous empirical results will be embedded in the context of the global city-regions. First, starting with the direct connection of the gaijin ghetto with the global city-region, it is followed by a discussion of the transitionists type, especially those who "lose their status" as a gaijin ghetto migrant (by losing their position within a transnational corporation or even their jobs in the industry), and how their downward mobility is connected to the global city-region. Then it closes with the proposal of distinguishing the Pro-Tokyoites—those venturing out from the limited spaces of the gaijin ghetto—as the surprising "pure" global city migrants with limited expansion and impact on the global city-region. At the end of the chapter, the debate is complemented by looking at the dynamics of spatial diversification from a migration industry point of view, juxtaposing the case of Filipino transnational migrants from below as an excursus, following more conventional routes of migrant industries known from the literature, to that of the novel dynamics of transnational corporate professionals' migration industry.

7.1 The Gaijin Ghettos into the Global City-Region

Although the gaijin ghetto in Tokyo is spatially limited within the city with its core of $4km^2$ in the Golden Triangle and a few more square kilometers when adding the central business district around Marunouchi or the work places at the waterfront, the description of this socio-spatial pattern already shows some diversions. On the one hand, there are residential exclaves in the upper-class suburban areas as noted before, especially inhabited by those foreigners who search either for proximity to the national schools, for example, French or German schools located in Yokohama, or desiring a house "*with a garden to barbecue*" (Real Estate Agent 04), that is, those with a cultural background affinity to a larger single family detached house than urban apartment living. In the context of these residential areas mostly inhabited by similar migrants and their limited social interactions with local Japanese communities—instead their focus is on school, company or consulate-based leisure and daily activities—these can be regarded as small exclaves of gaijin ghettos out of the global city area. Dots of gaijin ghetto exclaves can be observed in the global city-region, particularly in the intersectional areas between the Tokyo and Kanagawa prefectures, often on the verge of the outskirts of the central city or in one of the outer municipalities of the Tokyo prefecture. Clearly, the global city-region in which these gaijin ghettos venture out to is the western part of the region and limited to other directions where many "other" migrants, that is, Asian and lower-skilled migrants, work and reside.

On the other hand, the evolution of the gaijin ghetto itself can be regarded part of the dynamics within the global city-region where the global city area is being expanded to. As the Tokyoites noted, "*forests where tanuki*[5] *appear*" (Tokyoite 09) used to be found in the Roppongi area and it was not inhabited by the upper-class city-dwellers. The same can be stated about the waterfront area which used to be purely water and swamps where virtually nobody economically or socially interacted. Among the Tokyoites, the area around Tennozu is still an undiscovered area which they do not mentally grasp. Indeed, had it not been for the very specific constellation of these specific urban development projects focusing on these areas, they would never have become the hailed upper-class gaijin ghettos at all. Private urban developers and investors, such as the Mori Corp or the larger real estate developers successfully collaborated with the metropolitan government to push the agenda forward, and the unique involvement of the national government as part of the previously discussed developmental is the key to such urban dynamics.[6] The areas of the current gaijin ghetto are mostly those which have only recently come into being in the long history of Edo/Tokyo and are often still in development, stretching the boundaries of the city.

These two different aspects of the gaijin ghetto indicate not only the expansion of the city limits of the Tokyo global city in administrative terms but also in its overall social and economic extension. The boundaries of Tokyo have been dynamically shifting over the decades and centuries, both to the adjacent regions and also to inexistent spaces on the water. Although residents of gaijin ghetto exclaves may only have limited interactions with the neighboring population, their residence there impacts on the local economy and space in the global city-region.

Furthermore, when current economic policies and urban development plans are taken into consideration, it also becomes even clearer how the location and the overall expansion of the global city-space to the outskirts and eventually crossing the administrative urban borders into the global city-region increasingly unbind from solely urban actors. Sub-national,

[5] Japanese raccoon dogs; figuratively used to express remote, less inhabited areas in or near woods.

[6] It must be at least noted that these urban development projects might seem successful modes for upward revaluation of inner cities. However, such neoliberal urban development is rightfully highly critically discussed from the side of critical geographers. Reports are known about cases where urban developers pressured vulnerable older landlords to sell their lands to them with the help of dubious means. Issues of gentrification and extrusion of the original local population are important debates which are neglected in the neoliberal entrepreneurial city development in Tokyo.

that is, meso-level, actors of metropolitan and prefectural governments and also actors of the regional economy—such as private investors, private real estate developers and the otherwise global city-centered transnational corporations and their transnational professionals—contribute to a highly interwoven system of the global city-region. The interplay of such regional actors of the new regionalism, as discussed by regional scientists, is then embedded in different levels of governance, whereas in the case of the Tokyo metropolitan government in its paramount capital function of the country, the national government also has a say in the urban-regional-nexus by setting an agenda. Examples for such a national-regional collaboration are the plans to establish special economic zones. In contrast to the overall national migration policy, low-skilled labor migration is explicitly permitted in these zones. Moreover, collaborations between the national government and ministries with the metropolitan government especially for mega-events, such as the Tokyo Olympics 2020, aim for an overall economic boost for both the metropolitan region and the whole national economy. The newest trend in this context is the further development of the broader Tokyo Bay area, which will be one of the core places of the Olympic Games and also a tentative area of a further gaijin ghetto where a new economy is planned to be located.

7.2 The Downgrading of Transitionists out to the Global City-Region

Transitionists, as one of the three types identified within the transnational professionals, are those who show a stronger tendency to mobility or flexibility within the spectrum of socio-spatial patterns between the gaijin ghetto and the Pro-Tokyoites. Depending on the large variety of life course moments and decisions, the transitionists change their socio-spatial pattern to one direction or another, whereas there is no mono-causal relation to be identified regarding the reasons or the characteristics of the individuals. The family extension can contribute to a stronger attachment to the local in the sense of the Pro-Tokyoite in one person; in another person, the fact of having children can contribute to the reverse movement of going into gaijin ghetto patterns. However, as discussed previously, what remains common in the diversity of transnational professionals is the impact that capital can have in the end. The accumulation of monetary capital reflected in the income, seniority in position and also age tended to cause changes in the socio-spatial pattern of both gaijin ghetto and Pro-Tokyoite alike because of the changing availability of resources to define and decide on the socio-spatial patterns.

The transitionists take a special part in the debate on the dynamics in the global city-regions. They actually experience such a change in capital, yet not in the otherwise upward mobility of capital accumulation, but its loss through the economic downturn. In this particular case, the transnational professional, who had originally not been a financial professional but entered the transnational corporate world as a career changer from language teaching, lost his contract in the "expat package" because of the world financial crisis and was forced to change into a local contract with less income and fewer privileges associated with the expatriate benefits. At the time of the interview, concerns were voiced about maintaining the living standards of the gaijin ghetto lifestyle, including affording the rental for housing located in central parts of Tokyo and costs of the international schools of the children, and also the problem of a change in school systems to a local school and the future financial burden of higher education if the children choose to study in private universities in Japan or the United States.[7] The relocation was planned to be from the central gaijin ghetto area to what in Hall's words would be an "internal edge city" (Hall, 2011: 74) of Shinjunku in the furthest western part of the polycentric metropolitan region of Tokyo.

The interesting case of such transitionists, who start losing their status as upper-class transnational professionals thus becoming upper middle-class migrants, in a certain sense reflects the point that Sassen (2011) makes in her discourse on the difference between the global city and the global city-region. Although it is less the case of a positive connotation of equal distribution of capital for the middle class but more of a dramatic

[7] Due to Canadian citizenship, the transnational professional was taking a mid-term relocation back to Canada into consideration so as to increase the chances of his binational children to receive higher education without tuition fees.

change in the lives of transnational families falling victim to the global economic dynamics, the pattern of these transitionists can be still regarded as exemplary for the middle classes who are drawn out of the city center. Just as Fainstein (2011) discussed, as a consequence of the high living costs which they cannot afford anymore due to decreased monetary resources, these transitionists are forced to relocate, though not fully to the periphery but to the less central parts within the metropolitan area (Fig. 7.2). Simultaneously with the forced relocation comes the risk of also changing schools for the children, which in turn is successively connected to the change in the social communities and socio-spatial patterns. These dramatic situations—fortunately only one case in the interviewed transnational professionals—are indeed a reflection of what Fainstein pointed out as the phenomenon of "those global cities whose fortunes are particularly tied to financial markets" (Fainstein, 2011: 295) and by which the residents in them are vulnerable to these dynamics. Yet, what makes it an issue of not only the global city itself but the global city-region is the consequential socio-spatial mobility and the socio-economic downward mobility experienced by those who fall victim to economic downturn.

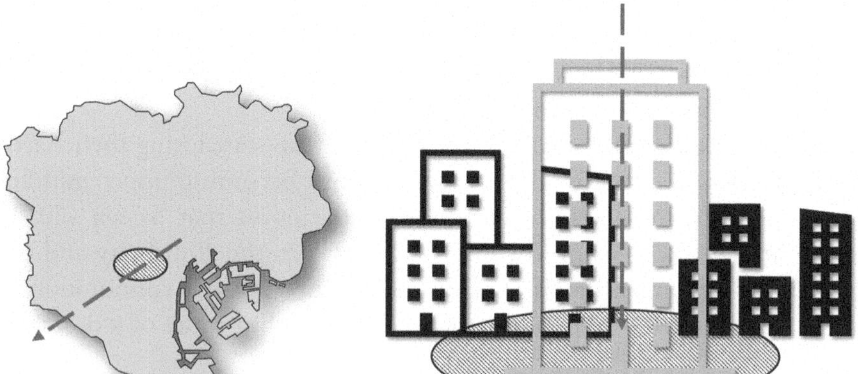

Fig. 7.2: Multidimensional shifts of transnational space (author's own)

7.3 The Pro-Tokyoites as Pure Global City Migrants

The Pro-Tokyoites—with a stronger attachment and similarity to the socio-spatial patterns of the "old money," specifically the old generation of Tokyoites—are characterized by socio-spatial mobility compared to the new money group of gaijin ghetto transnationals. In contrast to the gaijin ghetto pattern which is greatly centralized within the central parts of Tokyo and limited in its spatial scope, the Pro-Tokyoites can be regarded as those who are more "adventurous" and venture out to areas which gaijin ghetto people would not go to but where local Tokyoite peers tend to sojourn. Their transnational space within the urban, thus, is larger in spatial extent and also more diverse in the socio-spatial pattern as social interactions with the local Tokyoites occur more often.

However, what becomes interesting with regard to the connection of their socio-spatial pattern to the larger global city-region is that Pro-Tokyoites actually only venture out to the suburban, or more often the peri-urban, areas of the larger metropolitan region on a temporary sojourn. In line with Tokyoites' socio-spatial pattern of spending weekends outside of the core center for recreational purposes, such as in Hakone or Kamakura, the Pro-Tokyoites cover a longer distance out of the global city center, joining such outings, but only for short trips of tourism or leisure. The everyday practices surrounding work and residence remain in the city borders of the global city, making them less mobile on the regional scale and in the mid-term temporary context.

The dynamics of these Pro-Tokyoites signifies the importance of analyzing socio-spatial patterns at different levels and dimensions. Whereas within the urban borders, their space is much larger and more dynamic compared to their gaijin ghetto peers, their socio-spatial pattern in the regional scale is more limited and their social group can be characterized in its nature as being less mobile. Especially with regard to the temporal aspect of socio-spatial mobility, it becomes clear that there is a differentiation needed to distinguish the impacts that the different types of transnational professionals can have on the global city-region. Residential relocation and the expansion of whole work and residential areas into the global city-region, such as in the case of gaijin ghettos, undeniably show a stronger impact and connection to the city-region, whereas temporary outings of a private recreational nature are limited in the dynamics of the same region.

7.4 Excursus: Migration Industry from Below—Diversifying Filipino Migrants

Bringing the three dynamics of transnational professionals' connection to the global city-region into a more abstract debate, drawing out the mechanism of structures and agency in the migration industry in which they are embedded, this excursus will first present novel insights from what has been described as the "conventional migration industry" (Yamamura, 2022) of transnational migration from below. It will then reflect on the mechanism of the transnational migration from above, to holistically capture and analyze the mechanism by which transnational migrations of different types are coming together or diverging in the urban space. This excursus is taken from the publication "From global city-makers to global city-shapers: migration industries in the global city-networks," published in *Urban Studies* (Yamamura, 2022). The following is an excerpt on the empirical study with Filipino transnational migrants from below:

* * *

⁑ As interviews with recently arrived Filipino migrants show, the migration networks that recent migrants from the Philippines rely on and use as the source of information and support, are based mainly on family and friends, that is, stronger ties. Due to the offensive dispatchment of Filipino labor migrants throughout the world through the Philippine Overseas Employment Administration (POEA), those in Japan mostly have closer relatives and at least acquaintances in their social network who have emigrated, too. Though family members and friends might have moved to Canada or Australia, and might not always be present in

Japan, with the large number of Filipino overseas workers in these countries, the communities of co-ethnics are existent and also heavily used to share experiences and social capital before and upon arrival. As one of the interviewees who came through a language school, changed into employment in a Japanese firm and recently to a transnational company accounted his background and arrival:

> My aunt and my cousins are in Canada, and I have friends who went to Australia. They are everywhere. We are everywhere. That is the Filipino thing. (…) So, I got the job through my friend's sister who told me about the language school in Japan. They come to Philippines, a lot of them. And I went to school and came to Japan. I didn't know anybody here, my family and friends are in Canada and Australia, but the here are many Filipinos. Different people. (Interviewee B05)

As much as the information and connections within the co-ethnic migration networks are crucial for transnational migrants from below coming to Japan, the migration industry that—so to say—manages the migration channels is of considerable importance, too. As it is conventionally the case with unskilled and lower-skilled migration to higher economic countries, Filipino informants emphasized the role of brokers. Similar to the case of Vietnamese migrants in Japan (Muranaka, 2023), the migration industry of Filipinos consists of recruitment agencies cooperating with employers often in manufacturing or agricultural industry, even coupled with government registered intermediaries, who recruit, accompany and execute the visa application process (cf. Debonneville, 2021). As language fluency is crucial to the integration to the labor market and also a prerequisite to securing a stable visa, language schools both in the home country and in Japan upon arrival play a prominent role within this migration industry.

> I went to the Japanese language school in Manila first. There, I had to learn Japanese, complete first test so I could come to Japan. I was here as language student, and I started baito [translation: part-time job] in the company I got working visa later. The school took care of everything, the visa, getting job. (Interviewee B01)

It is through such language schools, the Filipino migrants report, that they first receive the student visa status, and also receive information and more importantly references to companies that will be employing and provided them with working permits upon completion of Japanese language courses. The role of international education as a novel migration industry has been pointed out by other authors on Chinese migrants to Japan, too (Liu-Farrer & Tran, 2019). It points at the necessity of this route especially because of the Japanese government's reluctance of opening doors to labor migration. Such information on recruitment agencies and language schools are shared within the co-ethnic communities, leading to a highly interwoven system of the migration industry with the migration network (as illustrated on the left-hand side of Fig. 7.3).

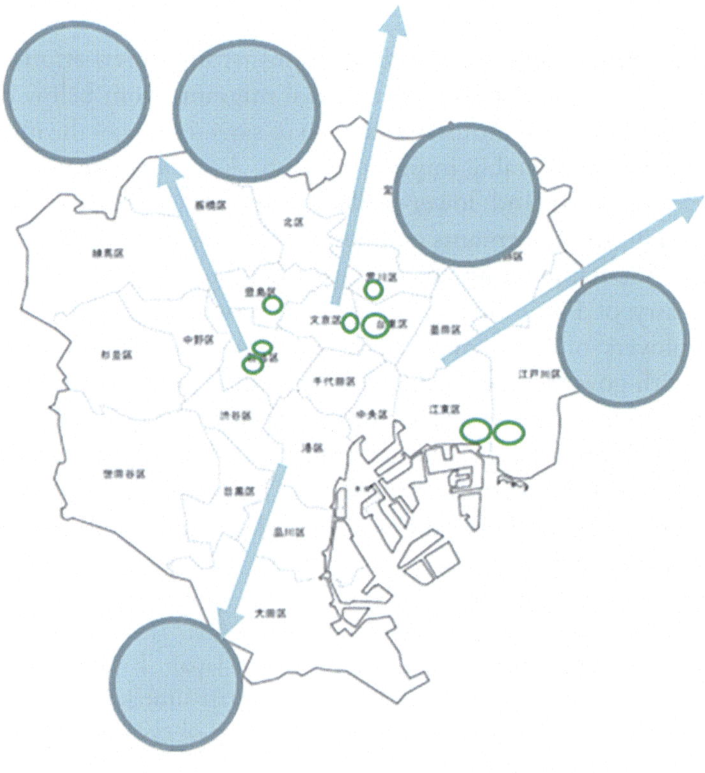

Fig. 7.3: Ethnic towns and the main border hubs of Filipino migrants from below (Yamamura, 2022: 2243/CC BY 4.0 DEED)

In the urban context, this particular linkage and embeddedness of migrants in the "conventional migration industry" results in specific socio-spatial patterns, which define the shaping of the city. The urban spaces of such migrants from below are strongly related to the location of the language school, where they are obliged to study during the student visa status, and more importantly also of the location of either the company housings when once employed or where co-ethnics reside. The companies cooperating with recruitment agencies decide primarily on the residential location and through that also the geographical extent of the everyday socio-spatial practices of these migrants. As one of the Filipino migrants explained:

> I'm not here as Genba [colloquial terminology signifying construction workers], but everybody who come here and work in factories or at genba [construction site], they all live far away. It is closer to where they work, they get there in busses and live in the dorms out there. The company give them housing and they get picked up and go to work. They don't have time to come to Tokyo so I don't see them in the week. On weekends, I go and meet them. (Interviewee B02)

As depicted in the map based on the interviews with Filipino migrants, the housings and the everyday life tend to be located at the outskirts of the Tokyo metropolitan area in logistic hubs (see Fig. 7.3). In these border cities, such as Omiya (as entry point to Saitama prefecture) or Funabashi city (the city in Chiba prefecture right at the border from Tokyo), with their function as major transportation hubs to the adjacent prefectures, housings are more affordable for the companies and also close by the work in the manufacturing and agricultural industries. a similar pattern of the outskirts can also be observed for care homes spread in the broader metropolitan area, where migrants would also be located.

The other points of interest for these migrants from below are the aforementioned language schools, which can be either located close by these border cities, too, or in more central locations, such as Shinjuku or Ikebukuro (Western edge cities of the Tokyo city center), and social meeting places. As one of the Filipino care worker migrants explained:

> In-between work and language school, there is nothing more. School is in Shinjuku, the company chose it, and we live here [in Adachi in Eastern part of Tokyo]. (…) Actually, the housing fee is already deducted from my salary because the company organizes it all. (PB05)

With such tightly scheduled weekly routines, there is not much time for further activities, yet shopping Filipino foods and socializing with other Filipinos are carried out in more central areas like in Ueno (Ameyoko area and Ueno park) or Catholic churches which partly even offer services in Filipino for the large community around Tokyo. What is interesting here to mention is the online space of social media which is crucial in sharing and receiving information not only for the migration process itself, but in finding the locations for the urban practices within the Tokyo global city-region as accessing the community market. The “shop under Ameyoko,” for example, has been mentioned as one of the main grocery shops where Filipino and other Southeast Asian food can be purchased, and is well propagated among these migrants by social media. When asked where he went for ethnic food-shopping, the Filipino migrant pointed out:

> Go to Ueno Ameyoko in the underground, there’s where everybody goes. (…) I found it [Asian food store run by Filipino] through Facebook and my friends told me about. We go there on weekend. (DB07) ⁑

* * *

As the excursus illustrates, the theoretical debates surrounding the global city-region (see Scott et al., 2011) resonate what has been developing indeed on the side of transnational migration from below. Transnational migrants from below rely on and are differently embedded in the mechanism of conventional migration industry compared to transnational professionals. They are dislocated to the outskirts of the metropolitan area, which is the global city-regions or to the gateways to the adjacent regions, but also restricted in their mobility due to their everyday constraints to venture out further.

In an interesting take of socio-spatial mobility patterns of transnational migrants in the Tokyo metropolitan region, the excursus shows how in both cases, transnationalism from above and below, migration industries are strongly contributing to the making and shaping of the global city and its region. While it is the recruitment agencies, language schools, complemented by strong social networks of family and co-ethnic friends which help transnational migrants from below to navigate the transnational migration itself and urban space, in the case of transnational professionals from above, as we have seen it in the previous chapters, are reliant on the corporate migration industry. The corporate (versus the conventional) migration industry consists of the HR departments of transnational corporations, the commissioned relocation companies and real estate agencies, with limited migration networks focused on business and co-worker communities. The different actors in the migration industry types contribute to each their specific socio-spatial dynamics, but the mechanisms of socio-spatial patterns produced and reproduced are interestingly similar.

7.5 The Global City-Region Dimension of Transnational Spaces

In response to the question posed at the beginning of this chapter, connections of the socio-spatial practices of transnational professions to the global city-region are clearly existent, and all three types are characterized by different dynamics. Whereas the transitionists' and the Pro-Tokyoites' involvements in the global city-region are mainly based on the dynamism of the transnational professionals' individual "ways of daily life" and more distantly related to their socio-economic context, the dynamism of the gaijin ghetto is clearly a case of stronger structurally embedded context.

The Pro-Tokyoites are those who, although expectedly being more prone to venture out of the limited transnational urban space of gaijin ghetto within the Tokyo global city, actually remain based in the core global city with only private and recreational temporary sojourns to the peri-urban outskirts of the metropolitan region, following the old

Tokyoites socio-spatial patterns. The case of the dynamic of the "downgraded" transitionists is interesting insofar as these are transitioning into a completely different socio-economic class and, by that, all-embracingly change their socio-spatial pattern along with other middle classes into the suburbs and the larger metropolitan region outside of central Tokyo. Yet, as they lose their status as part of the upper-class transnational professionals, this dynamic should be observed with care. Last, but not least, the gaijin ghetto who are—at least in the context of their socio-spatial pattern within the global city itself—less mobile and limited in their spatial extent are the ones who are most connected to the expansion of the transnational space beyond the city borders.

The expansion of the gaijin ghetto in the global city-region should be understood and also observed in its embeddedness in such larger events and strategic regional developments at different levels of policies, particularly economic, as outlined. The socio-spatial pattern and the locations of these gaijin ghettos are strongly embedded in the global, national, regional and urban arena of policy changes and characterized essentially by dynamism. Yet, what makes the transnational space in and around the Tokyo global city so specific is such diversity of dynamics of socio-spatial patterns in a complex system of influences at different levels beyond both urban and national borders. Having discussed the regional level of the phenomenon of spatial diversification in the global city context, the next step then is to scrutinize how the overall transnational space is constituted on the global scale, which is the overarching network of global cities.

References

Debonneville, J. (2021). An organizational approach to the Philippine migration industry: Recruiting, matching and tailoring migrant domestic workers. *Comparative Migration Studies, 9*(1), 1–20.

Fainstein, S. S. (2001). Inequality in global city-regions. *disP - The Planning Review, 37*(144), 20–25.

Fainstein, S. S. (2011). Inequality in global city-regions. In A. J. Scott (Ed.), *Global city-regions: Trends, theory, policy*. Oxford University Press.

Hall, P. (2011). Global city-regions in the twenty-first century. In A. J. Scott (Ed.), *Global city-regions: Trends, theory, policy*. Oxford University Press.

Hamnett, C. (1996). Social polarisation, economic restructuring and welfare state regimes. *Urban Studies, 33*(8), 1407–1430.

Liu-Farrer, G., & Tran, A. H. (2019). Bridging the institutional gaps: International education as a migration industry. *International Migration, 57*(3), 235–249.

Muranaka, A. (2023). 'Framing' and 'packaging' of foreign skilled workers: Diversity of the intermediary actors in the cross-border labour market between Japan and Vietnam. *Globalizations, 20*(4), 696–713. https://doi.org/10.1080/14747731.2023.2165376

Sassen, S. (1991). *The global city: New York, London, Tokyo*. Princeton.

Sassen, S. (2011). *Cities in a world economy*. Sage Publications.

Scott, A. J. (2011). *Global city-regions: Trends, theory, policy*. Oxford University Press. https://books.google.co.uk/books?id=mHE0b45cEEUC

Scott, A. J., Agnew, J., Soja, E. W., & Storper, M. (2011). Global city-regions. In A. J. Scott (Ed.), *Global city-regions: Trends, theory, policy*. Oxford University Press.

Sonobe, M., & Machimura, T. (1997). *Globalization effect or bubble effect? Social polarization in Tokyo*. Workshop on Global Cities, City University of New York' Graduate Center, New York.

White, J. W. (1998). Old wine, cracked bottle?: Tokyo, Paris, and the global city hypothesis. *Urban Affairs Review, 33*(4), 451–477. https://doi.org/10.1177/107808749803300401

Yamamura, S. (2019). Transnational professionals' socio-spatial venturing out to the Tokyo global city-region. *Regional Studies, Regional Science, 6*(1), 512–519. https://doi.org/10.1080/21681376.2019.1677173

Yamamura, S. (2022). From global city makers to global city-shapers: Migration industries in the global city networks. *Urban Studies, 59*(11), 2234–2254. https://doi.org/10.1177/00420980221087927

8

World at Home: Home in the World

Having presented the spatial diversification occurring in terms of the shaping of transnational spaces both within the global city and its expansion to the global city-region of Tokyo, the final step of the process of zooming out and recontextualizing the dynamics at different levels is to take a look at the global perspectives. This chapter encompasses three different pars: In a first step, we start with looking at the actual spatial perception of the transnational professionals based on the global mental mapping of the transnational space drawn by them and the narrative accompanying these maps. Then, secondly, these findings will be discussing these empirical results from the spatial theoretical lens, and finally, the chapter is followed by a more comprehensive conceptual debate on the spatial transformation of the global cities on a global scale. This chapter thus identifies the actual transnational nature of the space which is constituted by the global economy on the structural side and also produced by the socio-spatial practices of transnational professional on the practical side. In other words, it illustrates that the transnational space is not only translocal but also embedded in the transnational context of the global economic circumstances in which transnational corporations, and through them the transnational professionals, are embedded in.

S. Yamamura, *Spatial Diversity in the Global City*, Global Diversities,
https://doi.org/10.1007/978-3-031-64725-3_8

Returning back to the initial concept of the spatial theory under which the transnational space has been introduced at the beginning of the monograph, the analytical framework of the spacing and synthesis processes will also be applied to the global scale as in the discussion of the local space, yet with some abstracted transference. With the space being a "relational ordering of living entities and social goods" (Löw, 2008: 35), it is defined by the two analytical processes of the spacing and synthesis. Whereas the latter process can be easily depicted and deduced from the interviews on the actual transnational socio-spatial patterns, the spacing of allocating physical materials and the placing of oneself is more difficult to grasp on the global scale. However, allegorically, the placing of oneself and the location of the physicality can be interpreted as being directly reflected in the drawing of the specific areas, be it cities or regions, into a world map. The mental mapping of the world in which the transnational professionals place themselves is only difficult to separate from the synthesis processes of the global scale of the transnational space; as a matter of fact, for the local level the visualizations in mental maps were used for the analysis of the synthesis process instead of the spacing. Yet, for the global dimension of the transnational space, the analysis of spacing in the form of site surveys and ethnographic work is neither sensible nor practicable. As an approximation of the spacing, so the placing of specific physicalities into space, the mental maps can be used with the complementation of the naming of the areas in interviews to map out the transnational space in its spacing. The characteristics, usages and allocation of meanings to these physical spaces expressed in the mere city and regions drawn onto the map are then the actual reflection of the "process of perception, ideation and recall" (Löw, 2008: 35), that is, the synthesis process.

Basing on this transference of the spatial concept to the abstract level of the global, this chapter analyzes: (a) the spacing of the transnational space as depicted on the mental maps of the transnational professionals' world; and (b) the synthesis and its implication for the socio-spatial character of the transnational space deduced from the interview material. The conclusive discussion (c) then reflects the overall dispersion and expansion of the transnational space on a global level, taking into account its local urban and global city-regional spread within the network of global cities. A terminology which has been brought up by one of the interviewed transnational professionals in this context is what has been called Roppongization by one of the interviewees and which will be adopted here as a concept illustrating the emerging trend of global urban dynamics. It can be roughly translated into the global spread of universal transnational spaces of the gaijin ghetto. This concept will be further discussed and juxtaposed to the also globally universal, yet more locally anchored, spatial diversification of the Pro-Tokyoite space, which then is contextualized in the phenomenon of spatial superdiversification in global cities.

8.1 Spacing of the Transnational Space

The spacing depicted in the mental maps drawn by transnational professionals in the form of placing the locations of their transnational space on the global scale shows several patterns. Adopting a more systematic approach to analyzing the world maps rather than individually describing the maps, the following characteristics distinct to the interviewed person group and their socio-spatial patterns—and relevant for the understanding of their transnational spaces—can be identified.

Types of World Maps

Adhering to methodological correctness, the transnational migrants were asked to draw their mental maps of the world, without any specific constraints or suggestions of the graphic design of the map. The only factual constraint was the A4-paper sheet provided (though also multiple sheets were given when requested). Only on inquiry it was explained that not necessarily a "geographically correct" map was expected and that they were indeed free to choose the way to visualize. Three types of visualization evolved by this open approach: a network map with connecting lines between places (Fig. 8.1), an abstract map of island or circle-shaped areas (Fig. 8.2), and, most often, continents showing rough resemblances to common world maps with dots signifying the specific places of interest (Fig. 8.3).

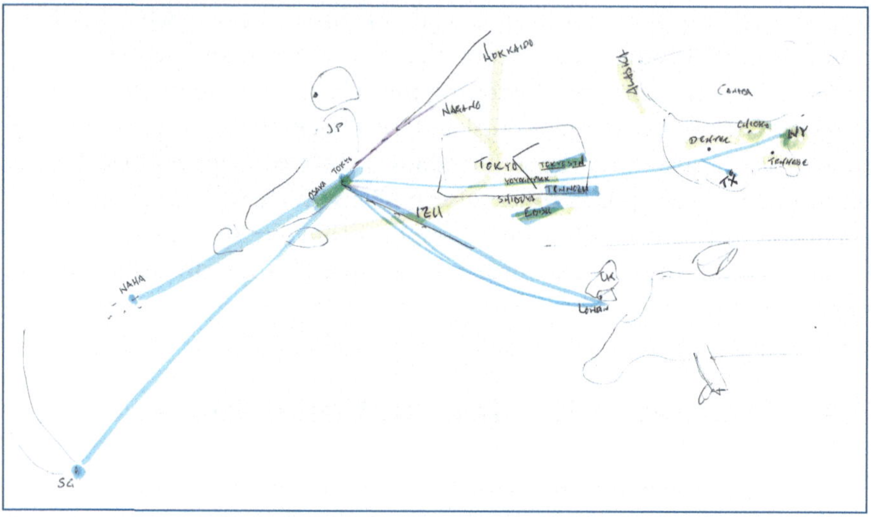

Fig. 8.1 World mental map—Network type (research material of author)

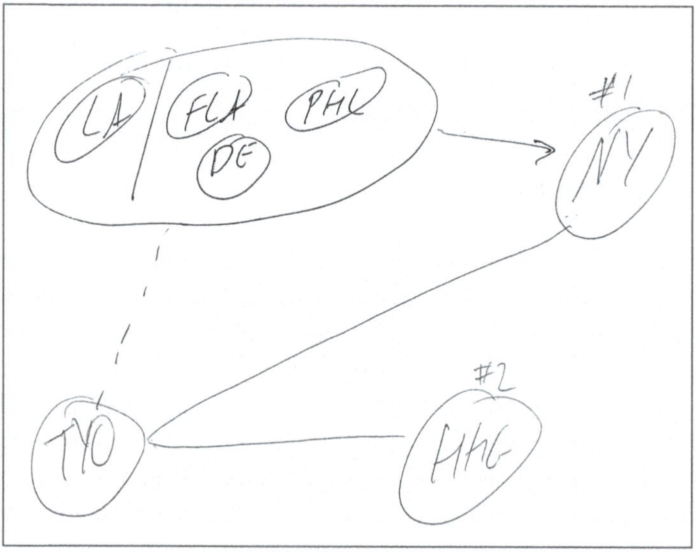

Fig. 8.2 World mental map—Abstract type (research material of author)

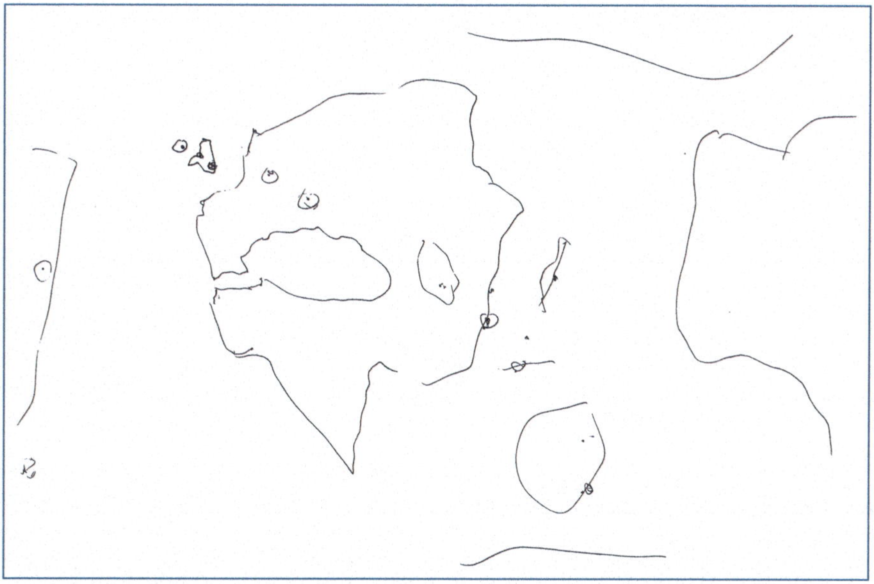

Fig. 8.3 World mental map—Geographic type (research material of author)

Inclusion of Elements

The elements drawn onto the maps are the direct transference of the places in the transnational space which constitute the world in which they live in. Apart from the different visualization types, what is common for the maps and also distinct for these mobile transnational corporate professionals is the clear selection of primarily global cities, in particular of the alpha category (Fig. 8.4). As banal as this may seem, in the sense that it is logical that these people who are financial professionals in transnational corporations are business-wise focused on the main financial centers of the global economy, and obliged for transnational hypermobility for short-term assignments, business trips and relocations to a specific global city, it is still the distinct character of their transnational space which is different from transnational migrants in other professions, especially those who are moving back and forth between specific sub-national regions. What also becomes clear is that the city-focused mobility and interactions are reflected in their private social ties. With the exception of

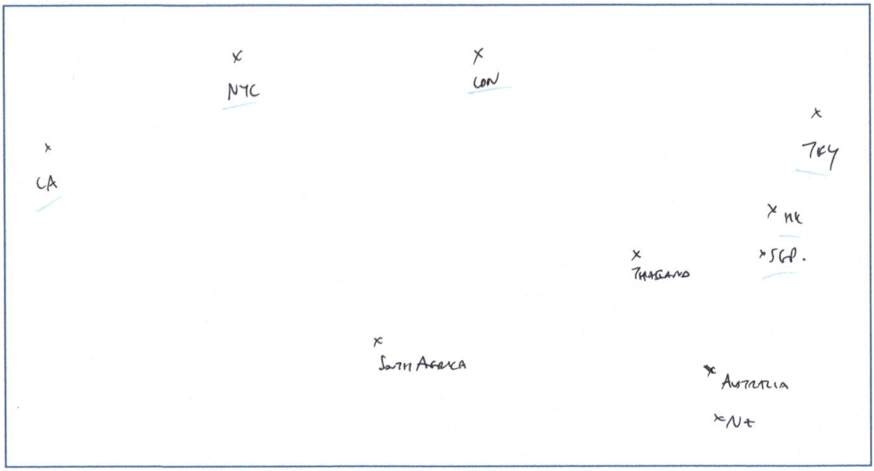

Fig. 8.4 Global cities centered world mental map (research material of author)

a few diversions illustrating places of leisure or even resort spots and individual locations depending on the ethnic and other family backgrounds, an overlapping of business-related and personal spaces is observable. Conceptually, there is a clear indication that these subjective mental maps of the world by transnational professionals actually parallel the world mapped out by macro-level data on the network of global cities by the Globalization and World Cities (GaWC) researchers (Fig. 8.5). The GaWC maps are, therefore, not just the result of a conceptual practice of scientists, but a real phenomenon among these specific transnational migrants. The following juxtaposition emphasizes the overlapping between the conceptual construct and the spacing of the transnational space of the transmigrants:

Omission of Elements

An interesting aspect of the mental maps is not only what and how places are visualized, but also what is not put in. The omission of elements is in fact attached particular importance in qualitative research and, in the case of these mental maps, the omissions or the absence of elements can make

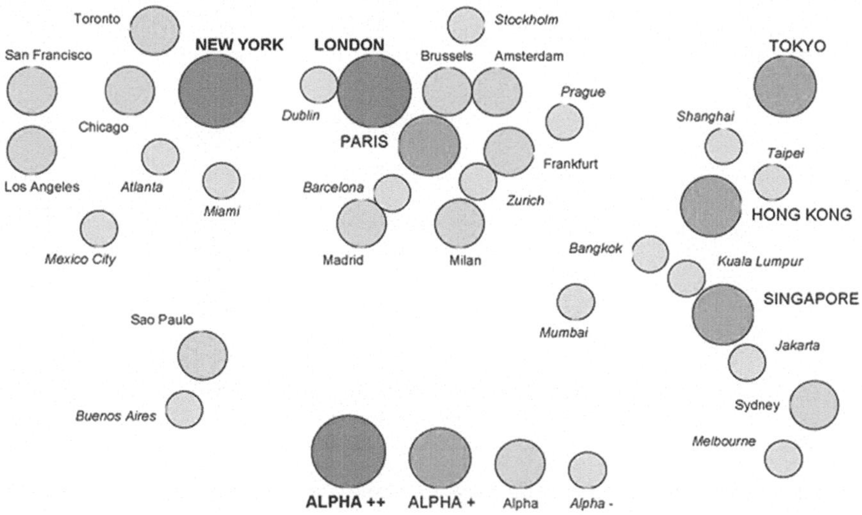

Fig. 8.5 The world according to GaWC—World Cities 2010 (GaWC, 2020)

the maps distinct for the socio-spatial pattern of people and reflect their world. Two clear characteristics can be mentioned in this context, namely, the omissions of continents and regions that are not relevant in their world and the disappearance of national borders (Figs. 8.6 and 8.7). Both are distinct characteristics of the transnational space of the interviewed professionals. This is again, on the one hand, embedded in their economic activity that continents or regions, such as Africa, Oceania and South America, are often left out, unless privately connected. On the other hand, whereas commonly borders are typical in mental mapping, the contours of countries are rarely found in the world maps of the transnational professionals, who yet again can be regarded as being business-wise but by that also personally focused on the global cities.

The latter trait of the transnational space stresses the actual nature of the transnationality of these people's space. Literally, their world crosses borders and, as can be seen in the two examples above, the Eurasian continent and the North American land mass are not illustrating any national

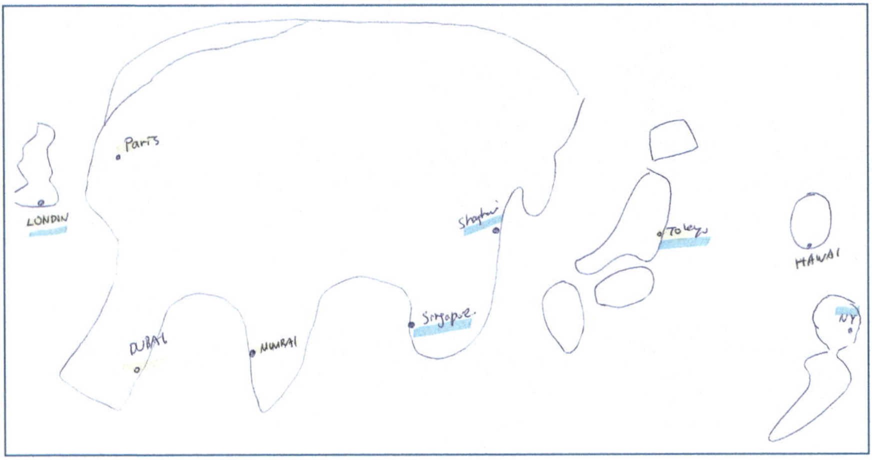

Fig. 8.6 Borderless world mental map, example 1 (research material of author)

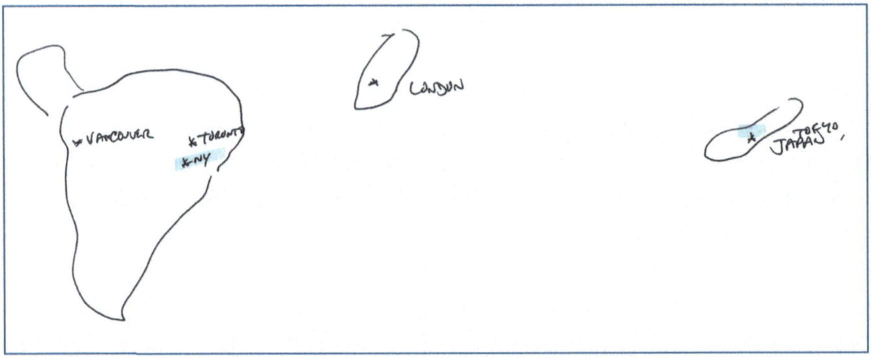

Fig. 8.7 Borderless world mental map, example 2 (research material of author)

borders. Island countries and those in the discursive center of the interview, for example, the UK or Japan, may be drawn in individually, but the overall tendency is for the omission of bordered nation-states. Instead, they function merely as the shape of the overall area, which can in more extreme cases, as in the figure above, be reduced to only the relevant cities and lose any bordering of countries or shapes of regions. Clearly, not every map of the transnational professionals looks the same and clearly not all elements are drawn in the same, yet the tendency for the transnationality of their space is undeniable and can also be detected in the following analysis of the synthesis process.

8.2 Synthesis of the Transnational Space

While drawing the world maps, the transnational professional migrants also started explaining and giving insights into their understanding of the maps, reflecting on their own behavioral patterns as well as their cognitive relation to them. These "imaginative, perceptive, and reminiscential processes" (Löw, 2001: 225 [author's translation]) shed new light on how the transnational spaces located in the previous section on the spacing are being created in practical terms.

One of the most prominent aspects of the emergence of transnational practices is the development and dispersion of internet and communication media. They enable and enhance the transnational social practices between migrants to an extent that was not possible a few decades ago, whereas these devices contribute to a different perception of, as well as practice with, space itself. The usage of digital media for keeping the transnational connections with one's home society and also friendships elsewhere has been much discussed in literature on transnationalism. As classical in transnationalism research, several transnational professionals mentioned the usage of Skype to keep in touch with families as a crucial element of their daily lives, clearly demonstrating the transnationalism of these migrants. Though, what is interesting for the discussion of the space-making are the changing ideas of space, as can be seen here:

> My home is definitely where my family is. So it's definitely still America, I call that my home. Ehm, I just have more than one home, I think…'cause I call London my home, too. [laughing] Ehm, I mean places become your home, where you fit in and where you find friends and where you feel comfortable. So, ehm, yeah … I miss … I realize that I am sacrificing being around my family, ehm, but I participate a lot in their life on Skype. Skype, I think, is this fabulous tool where you can feel like you're in your kitchen.

> They pop me up at family parties on the computer, people come by, say hi... (RF07)

As this example illustrates, it becomes clear that the transnational space-making is not limited to the socio-spatial pattern locally in the urban environment, but goes beyond the national borders, thus, becoming transnational in the actual sense.

The spatial implication of such transnational practices for the cognitive production of transnational space is even better illustrated by the accounts of their perception of spatial issues. On the one hand, transnational professionals often mention the focus on social ties which they keep beyond the borders rather than settling or identifying themselves with the geographical spaces:

> Because, [...] you know, I have friends, like, once you make friends it's, like, you know, you have got that connection to a place, I mean, that is what really encourages you more to a place, you feel more comfortable, I think. (RPL06)

> I, I think they [the transnationally mobile professionals] don't feel ... or they are not physically very much locked into a certain place, so they, they are very mobile but nevertheless are well connected both with their home country as well as with friends in, in other places of the world. I guess the whole, whole net- network of people is constantly, not constantly, but is on the move, but the connections are normally not lost. They stay up. I think that is what I would roughly describe. (GC16)

On the other hand, the fluidity or dynamics of the constantly changing spatiality of their everyday lives is another motif that appears in the narratives of the transnational professionals, whereas the lifestyle is clearly referred to in the professional context:

> I consider Japan more home than anywhere else. But right now, yeah, I mean, for, for the best part of the last fifteen or sixteen years, I've, I've really been based in APAC. So, really started out in Singapore and for the first two years, really, I lived in a hotel and traveled around the region. Most of it, again, covering for national communities, for ... and, and I was really a transient, I literally, I lived in a hotel and my boss would call me up and

> they would say: (…) 'We have a project up in Hong Kong. Go up there, do the project, when you're finished, we'll give you your next assignment and you can decide to either stay in Hong Kong or, or go back.' (BA17)

> And, you know, I, I'll challenge my family, and if we need to move and go with the work, we need to do that. (PL09)

In stark contrast to other transnational migrants who live and go back and forth between two countries, and also the diasporic communities, many of the transnational professionals report the loss of the feeling of home. Diasporic communities maintain, and are characterized by, definition through their common ideas of home and the other transmigrants, although reportedly having issues of being here and there or even in-between the respective countries, still tend to have a stronger attachment to the home culture and regions. The group of transnational professional migrants, though, often keeps their original attachment to their home country or citizenship, but increasingly loses the spatial attachment to their home.

> Do I feel homesick, do I feel homeless? Yeah, I guess I do because I don't maintain a permanent… I sold my property in the UK, and as I mentioned I sold my house in Japan. Yeah, I don't have a base per se. I guess my parents in the UK would be the most fixed address I could ever give within any two-year period. (C13)

> Me, yeah, I feel like a bit of a loner myself, to be honest, because especially now that I have left Japan, which was my base, you know, it was my…my, what would you say? my adopted base after leaving the UK. And, I cannot consider the UK anymore. It's somewhere I was born and spent the first half of my, my life. But the times I've been back there, I just can't wait to get out again. So, yeah, if anything, I'm an Asian citizen and it's my adopted region, if you like. But yeah, being transient, I mean, (…) I've been thinking over it myself is that I don't really have a home anymore. So I'd probably be one of these global citizens and go where the work is. (LL29)

The examples are taken from different types of transnational professionals, yet the traits are present irrespective of the typified socio-spatial patterns. With these transnational professionals dispersing globally in specific cities, the question then arises as to what spatial implications their spatial perception and imagination have on the overall transnational space.

The spatial implications on the global level are twofold. One derives from the new lifestyle and patterns of the gaijin ghetto which, in accordance with the previously mentioned trend of increasing spatial mobility and also temporal dynamics of the transnational professionals, spread throughout the network of global cities. The other side is the increase also in the Pro-Tokyoite spaces within the transnational professionals' socio-spatial patterns. Both are occurring within the network of global cities as part of the transnational spaces and it can be asserted to be a universal trend of the expansion of these transnational spaces, but the nature of the spaces are essentially different and their significance for the spatial diversification of the global cities divergent. The following section discusses the so-called global Roppongization of the expansion of the gaijin ghetto-type of transnational space, its overall embeddedness in the global economy, and then the Pro-Tokyoite spaces which are contextualized more strongly in the notion of superdiversity.

8.3 Global Roppongization as the Emerging Trend of Transnational Space

The gaijin ghetto that has been elaborated in the previous chapters has probably come as little surprise because the phenomenon is commonly well known from any global professional or business expatriates. Mass media, non-fiction authors and academics alike have been reporting on these "expats" or "global elites." Yet, the socio-spatial implications of their social behavior have not been explored thoroughly. One strand of literature, though, that has discussed such global spatial phenomenon of places which grow similar in the end and are supposed to lose meanings as places is Marc Augè's non-places. His work on so-called non-places is best described and reflected in the words of one of the interviewees with gaijin ghetto patterns:

> Well, just because if you live, like, in most business hotels there, a template, and they offer very much the same thing. I mean, that's the business model, like, India is an emerging market but they have Hilton's, they have Grand Hyatt's and everything, same stuff. […] And so that is generally where I'd be and you look at the room—they look the same. Menus, I mean, maybe they've got a little bit of the local food but generally you have the same basic Westernized stuff on there, plus local so it's pretty … you can get pretty lost and not get a sense of where you are. (IN04)

Though Augè mostly refers to transportation means and places, such as airports and motorways, and his theory has indeed been criticized regarding the sense-making because, even as identical as these places may appear, they can still symbolize and represent specific socio-cultural values and provide the space for specific social practices bound to them, there is no denying that the supermodernity or simply globalization is creating, or at least contributing, to the creation of such spaces which are commonly and increasingly used by highly mobile professionals.

The gaijin ghetto can be regarded part of this trend of supermodernity in which professionals, like those interviewees of transnational financial corporations, find within both Tokyo and the respective global cities on their business trips and other assignments their spaces with increasing similarity and—more importantly with respect to their usually limited time and resources—convenience. The dispersion of the same businesses and the same services has been described by one of interviewees even as a "Roppongization," which is a neologism to express the reproduction of places like Roppongi—one of the core gaijin ghetto places—in other global cities:

> And yeah, it's true that Ebisu or Roppongi, what I can see, is quite similar to what I see in Singapore. (…) Maybe they [Jakarta] will try to copy what they see in Singapore. But now, it is the center of Jakarta: some of the shopping malls or the office buildings, all these restaurants, they are quite similar. Even this Roppongization gets through but that takes much longer time. (MM09)

The overall context of the production of such gaijin ghetto transnational spaces lies in the dynamics of the global economy, in which transnational corporate strategies and, as part of them, transnational professionals are embedded. As already noted previously, mobility in a restricted time context is typical for the industry, as well as the dependence of these migrants to the motions of the global economy. The constant movement of capital and by that also the transnational corporations' economic activity and consequently of the transnational migrants as executive actors of the global economy is best illustrated by the following quote:

> Most capital today is just moving, it doesn't build. And most of the movement of money is, you know, derivatives and money to make money to make money, it's not to build, build things, right? And so, when you look at a graph of the Morgan Stanley annual report today, it just shows these graphs of, like, you know, people making more money, right? Or, like, computer screens… So, financial services is about money, so most people of financial services move to where the money is. So, like, why do people rob banks? That's where the money is. You move to where the money is, right. (EM34)

How much the speed and frequency of migration, and by that the attached transnational socio-spatial patterns, are dependent on the global flows of capital becomes more than clear. Also, the dynamics of the global economy in the form of shifting foci of the global cities can be observed in the transnational professionals. Tokyo, though still one of the primary global cities, has been losing its economic power, especially as other Asian global cities have been emerging. In the course of this global economic development, shifts of not only flows of capitals but also transnational migrants and with them their social networks—and, thus, the transnational space—occurred:

> Virtually all the expats had left. All of my friends had dwindled away until, finally, it was my turn to go. (MG04)

Whereas the overall locations of the gaijin ghettos are constantly on the move on the global scale and also as the discussions on global city-regions have shown shifts within the metropolitan regional development and thus shown strong dynamics, the nature of the gaijin ghetto is actually rather constant. The gaijin ghettos can be found in virtually any global city where transnational professionals, also named as expats or globals, sojourn and with globalization contribute to a kind of global socio-spatial convergence of the global cities, namely, the effect of Roppongization. What Roppongi is for Tokyo, it is the Holland Village or Orchard in Singapore. Interestingly enough, this convergence itself is already diverse in that these transnational spaces are produced socio-economically by a homogenous transnational professionals' class with its own lifestyle and culture of globalism, yet this social group itself is increasingly diverse in its social constitution of ethnic and national backgrounds. This homogenization among professionals of multicultural and multiethnic backgrounds is in line with recent transnational corporate strategies aimed at creating a global mindset ("universalist perspective of HRM").

Graphically, the spread of such universal gaijin ghetto areas in the network of global cities can be depicted as in Fig. 8.8. While the gaijin ghettos might be spreading in different areas within a city, as can be seen in Tokyo, with the areas around the Golden Triangle but also with hubs outside, the characteristics of these ghettos remain constant, or comparable as a bubble within the cities. The convergences of the gaijin ghettos can be observed throughout the network of global cities, while the transnational community itself is not ethnically bound and in that respective characterized by high diversity (different shapes in the bubble).

8.4 Pro-Tokyoization and Superdiversity

The other side of the spectrum of the transnational socio-spatial patterns is fairly different from these high-speed and constantly dynamic, yet characteristically universal, gaijin ghettos. As discussed in previous chapters, the socio-spatial characteristics of Pro-Tokyoite transnational spaces consist of the proximity to the local upper-class socio-spatial patterns, yet also the overlapping with that of the gaijin ghetto. The overall attitudes of Pro-Tokyoite transnational migrants tend to be different and the social interactions and building of networks include a larger diversity of people, in particular local non-migrants but also transnational migrants of the local national backgrounds.

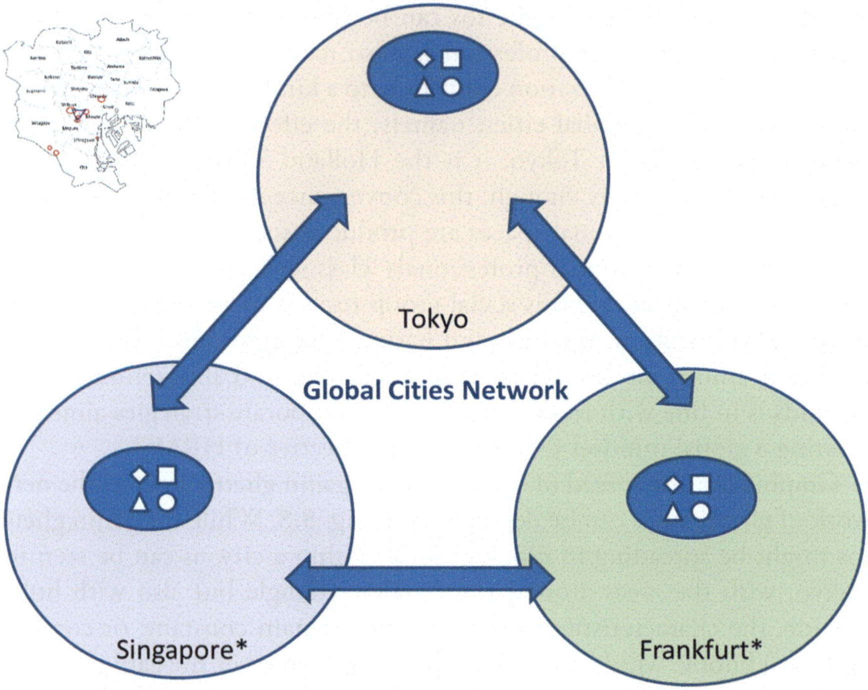

Fig. 8.8 Universal convergence of gaijin ghettos in global cities (Yamamura, 2022: 3394/CC BY 4.0 DEED)

The diversity of people is also reflected in the even stronger overall diversification of transnational space on a global scale (Fig. 8.9). The Pro-Tokyoites not only intermingle with the local peer population Japanese in Tokyo (thus a more diversified and expanded, even porous bubble within the global cities), for example, but when they move along to other global cities, which they do as much as other transnational professionals of the gaijin ghetto type, they also start to bind ties to the local people there. Thus, Pro-Tokyoite as a terminology becomes a mere archetype of the universal trend of transnational professionals merging with the local peer group, producing the transnational spaces through their social practices, yet with uniquely local characteristics to them. So, the Pro-Tokyoite will also become a Pro-Parisian, Pro–New Yorker, Pro–Hong Kongese, etc., wherever they move within the global cities network.

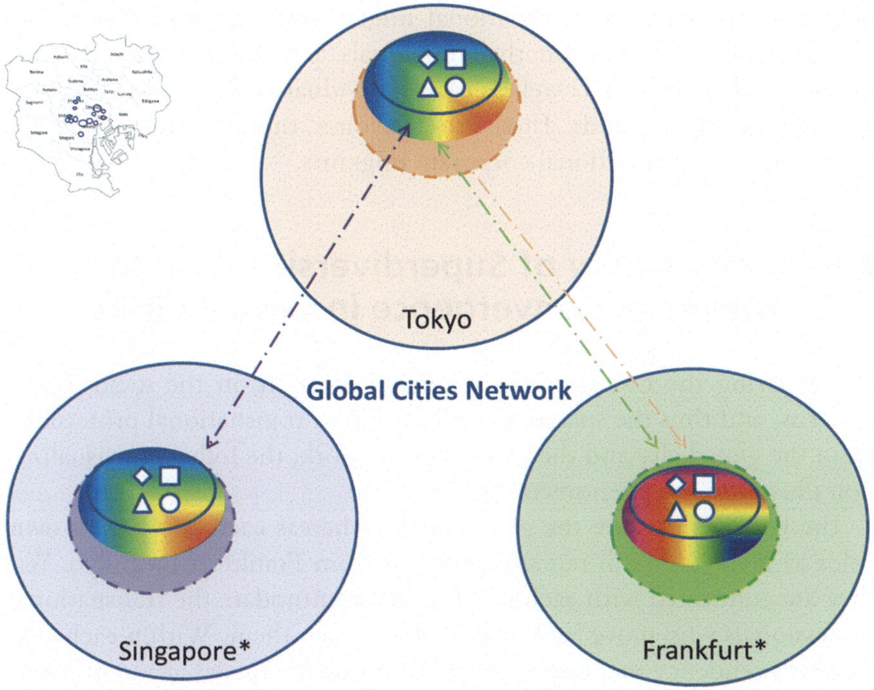

Fig. 8.9 Universality of superdiversification of pro-Tokyoites (Yamamura, 2022: 3394/CC BY 4.0 DEED)

Although the dynamics of the transnational space of the Pro-Tokyoite are rather constant in terms of the upper-class lifestyles, the accumulation of these Pro-Global-Cityites, so to say, brings about a group of transnational migrants each with multilayered and highly diverse socio-spatial patterns. The Pro-Tokyoites themselves are a superdiverse group of transmigrants with each being a hybrid in that their socio-spatial patterns overlap with the gaijin ghetto, especially due to their involvement in the transnational business communities and their corporate context. They, however, also venture out to the more local upper class, not adopting the universalist view but instead incorporating local cultures and socio-spatial practices. At the same time, the group of these transnational migrants, though similar in their Pro-Tokyoite attitudes and behaviors, is still diverse in itself as the multiple Pro-Global-Cityite characters accumulate with the experience of transnational migrations and also differ in the constellation depending on the individuals' professional and personal migration destinations as well as each individual's own ethnic, religious and cultural backgrounds. These Pro-Tokyoites, thus, illustrate the diversity within the transnational corporate migrants.

8.5 Universality of Superdiversification Vs. Universal Convergence in Global Cities

Summarizing the results of the empirical research on the socio-spatial patterns, and thus the spatial diversification, of transnational professionals in the global city and the global city network, the following visualization illustrates the phenomenon (Fig. 8.10):

The larger circles are the global cities, whereas each city has its own color as New York still remains different from Frankfurt or Tokyo. Yet, they are connected with each other and accommodate the transnational professionals who move back and forth between them. Within each city, there is a tendency to a gaijin ghettoization with a mixed group of transnational migrants (signified by the square, triangle, etc. in the smaller circle) who are spatially but also socially limited in the extent and, despite the multicultural backgrounds, these ghettos remain globally universal (and thus, the same small circles within all of the global cities). This convergence of gaijin ghettos over the global cities network is the process that has been named Roppongization before. On the other hand, the socio-spatial patterns of another group of transnational professionals overlap these gaijin ghettos, but at the same time also have a perforated circle to the local culture and space. This pattern of Pro-Tokyoite being hybrid in the sense of gaijin ghetto and each Pro-global city is illustrated by the same perforation of the larger circle. The color of these Pro-Global-Cityite circles, though, is dependent on each of the city because the "local" differs in each respective global city. The rainbow-colored parts within these Pro-Global-Cityite circles show the mixing of the diverse backgrounds of Pro-Tokyoites who can have different experiences and relations to other global cities by previous migrations.

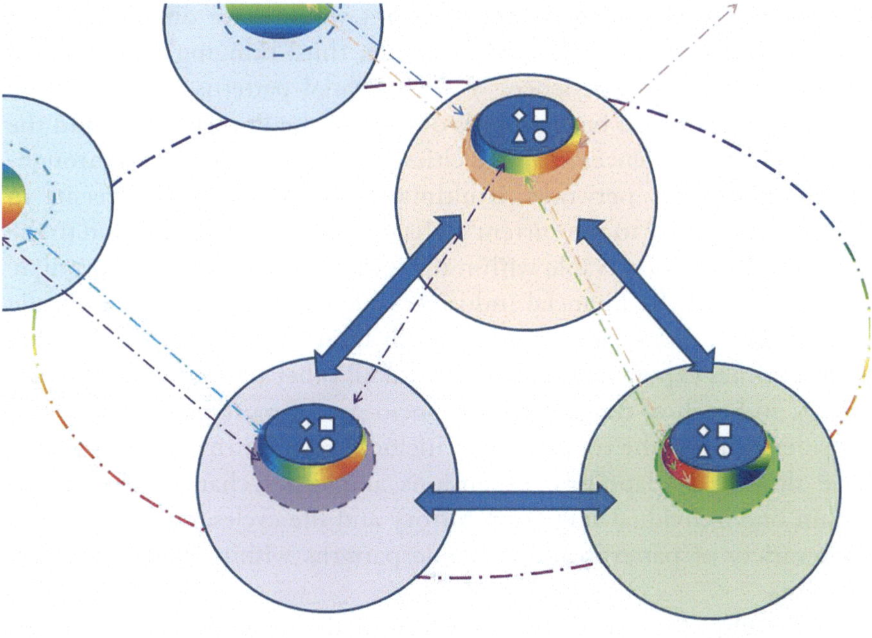

Fig. 8.10 Universality of superdiversification of pro-Tokyoites vs universalconvergence of gaijin ghettos (author's own)

Through the universal tendency to diversifying, that is, localize and venture out of gaijin ghettos, but also the diversity of migrant experiences accumulating within the group of Pro-Tokyoites, there is a superdiversity observable which is signified by the rainbow. Reflected on the global scale, the gaijin ghettos can be illustrated as more or less and probably increasingly universally convergent circles connected directly to each other in the global city network. In contrast, the Pro-Tokyoite—or symbolically the Global Cityite—is showing multidirectional ties of multidimensional diversity within the cities and a universality of diverse Pro-Tokyoite socio-spatial patterns on the globe.

The phenomenon of the Pro-Tokyoite appears to be universally spreading in society, especially in the context of the transnational space spanning the network of global cities. At the same time, the multiple layering of Pro-Global-Cityite socio-spatial patterns contributes to an even stronger socio-spatial diversification in these cities. The multi-layering refers to the fact that each and every transmigrant would have different sets of global cities they have migrated to before, thus, bringing a diversity of Pro-Global-Cityite experiences of socio-spatial patterns each. With all the different personal backgrounds of each of the Pro-Tokyoites and the multitude of experiences in global cities and the social networks throughout the global city networks, it ultimately creates a spatial diversity in global cities unique to the current global migration era. In addition to the diversity already observable within this group of transnational corporate professionals of the financial industry, which as the socio-economically elite of the global society is marginal in total population, the diversity even becomes exponentially diversified if all other transnational migrant groups and each of the sub-groups' socio-spatial patterns are taken into consideration. As the transitionists, including the downward movement, have shown, overlappings, expansions and other changes even occur within one individual migration history and life cycles, equating with a high variety of patterns and dynamic patterns within all transnational migrants.

In the end, the overall phenomenon of the superdiversity of spatial diversification is that through the multiple and frequent transnational migrations and mobilities of these people themselves, but also through the intersections of transnational spaces of an even larger diversity of other international and transnational migrants in these cities, the global cities as hubs of the global economy can be said to experience an almost exponential diversification. Features of social practices from other urban contexts of the Pro-Global-Cityites start to mix up as the ethnically and culturally diverse group of transnational professionals diversify even more, leading to a pure superdiversification of these diverse global cities. Especially in the long run, with further generations of transnational migrants growing up in these transnational spaces—such as the third culture kids, that is, children of mobile transnational families who, through the interactions with peer transnational children, develop their own social practices different from their parents' cultures and socio-spatial patterns—the multidimensional superdiversification of transnational spaces, in particular spanning the increasing number of global cities, will progress even more.

References

GaWC. (2020). *The world according to GaWC. Globalization and world cities research network*. Retrieved from September 15, 2023, https://www.lboro.ac.uk/microsites/geography/gawc/gawcworlds.html

Löw, M. (2001). *Raumsoziologie (Suhrkamp Taschenbuch Wissenschaft)*. Suhrkamp.

Löw, M. (2008). The constitution of space: The structuration of spaces through the simultaneity of effect and perception. *European Journal of Social Theory, 11*(1), 25–49.

9

Spatial Diversity in the Global City: The Spatiality of Superdiversity

9.1 Relationship of Migrants *and* Cities: Migrant-Led Diversity of Urban Spaces

Returning to the initial quote of the Introduction by Glick-Schiller and Çağlar (2011), that urban and transnational migration studies have focused much on the migrants *in* the cities as well as migration *to* cities, this chapter tries to fill, at least partly, the void on the nexus of "migration *and* cities" by reflecting on the spatiality of superdiversity. Developing from the idea of transnational urbanism by Smith (2001), several researchers have analyzed the specific context of migration in the urban context, with particularly Yeoh and other authors bridging the gap on the nexus of global or globalizing cities and transnational migration (Yeoh et al., 2000; Yeoh & Chang, 2001). Yet, as Glick-Schiller and Çağlar (2011) point out, conceptual debates on the relationship between cities and migration still remain underexplored and requite further attention. This is exactly where this monograph with its novel take on the concept of superdiversity and its focus on the spatial diversity in the global city crucially closes the gap.

S. Yamamura, *Spatial Diversity in the Global City*, Global Diversities,
https://doi.org/10.1007/978-3-031-64725-3_9

Superdiversity is a term to describe the multifactorial diversity of diversity in society, deriving explicitly from the increase of global migration (Vertovec, 2007). Indeed, the diversities of migrants are occurring not only regarding the ethnic and national origins, or further individual socio-cultural and socio-economic characteristics. The diversification also covers aspects of migration channels, legal statuses as part of migration policy contexts or migration networks through and by which individuals migrate. Superdiversity from the migration context can be understood to encompass the different forms and types of transnational migration. Moreover, the migration-led diversity in society can be found in different areas of the world. In fact, the existence and proliferation of different transnational migrant groups contribute clearly to the superdiversity in specific cities, also partly known as superdiverse cities or neighborhoods. The superdiversity perspective would then be not only to call out the ethnic groupings of transnational migration, but also to point at the mingling of different transnational communities with each other, and the diversification of the transnational communities.

However, while urban diversity has become an increasingly important not only for urban planning and design, but also within the particular research field of superdiversity, the spatiality of superdiversity in its actual spatial theoretical perspective has been neglected. Researchers on urban diversity discuss particularly issues of conviviality and social encounters in such highly mixed cities or superdiverse neighborhoods (Heil, 2020; Vincent et al., 2017; Wessendorf, 2013, 2014; Wilson, 2017; Ye, 2016) and even issues of "everyday multiculturalism" (Wise & Velayutham, 2014). This well-established thread of research is mostly focused on the challenges and opportunities that are connected to social encounters of people of different backgrounds. The spatial perspective is on the urban places in which such encounters happen, whereas the main focus remains on the encounters rather than the spaces. The current research primarily looks at the urban diversity issue in the sense of social spaces in which conviviality and encounters are emplaced (Berg & Sigona, 2016; Wessendorf, 2014) and the discourse on the urban spaces of superdiversity concentrates on such common places of encounters. As the fruitful concept of "commonplace diversity" (Wessendorf, 2013) points at, high levels of conviviality and inclusivity in urban space can be constituted to a degree that "ethnic, religious and linguistic diversity being experienced as a normal part of social life and not as something particularly special" (Wessendorf, 2013: 407). While it is suggested that "social spaces [...] play an important role in the process of familiarization with people who are different and in getting accustomed to communicating across difference" (Wessendorf, 2013: 410), the urban spaces and the mechanism of differential inclusion in cities remain abstract and the actual "spatiality of superdiversity" (Yamamura, 2022a) under-researched.

As the present monograph with its extensive investigation into the spatial diversity in the global city Tokyo and beyond has shown, the localization of such increase in diversities of socio-cultural interactions and social practices of transnationalism, especially in specific urban spaces on the globe, should be understood and further theorized as the socio-spatial expression of superdiversity. With the spatial perspective to superdiversity, the concept of superdiversity would be integrated more into the transnationalism research and its quest on the nexus of "migration *and* cities."

9.2 The Geography of Spatial Superdiversity: Inter-scalar Embeddedness

Empirical research so far has shown how—beyond the statistical analysis of migration data—society is diversifying in its urban contexts. Notable areas of research with a spatial component particularly within superdiversity research are the linguistic landscapes and anthropological approaches of trans-ethnography. Such results in superdiversity research illustrate how diversities in its different dimensions of linguistic, ethnic or even religious aspects become visible in urban space (Blommaert, 2013; Burchardt & Becci, 2016; Hall, 2015). These changes in the urban landscape have also been identified and systematically surveyed from entrepreneurial perspective. Entrepreneurial superdiversity, as the authors point out, is not only an issue of ethnicity or language, but also encompasses the diversification of ethnic minority and migrant entrepreneurship in terms of business types and also business strategies (Yamamura &

Lassalle, 2020). Furthermore, adding to these more migration-oriented entrepreneurial diversities, the intersection of gender minorities and migrants has also been recently studied in cities such as Amsterdam and Hamburg (Yamamura et al., 2022). These empirical evidences clearly demonstrate that superdiversity is not found simply everywhere in cities that are known to be hubs of global migration. There are subtle differentiations of the spatial dimension needed even within the superdiverse cities at a neighborhood or even street-level. In fact, what the latter research alludes at is the distinct and crucial socio-spatial context at the different geographical levels, in which persons especially at the intersection of diversity dimensions, such as LGBT* minorities who are also migrants and/or even people of color (POC), would be bound to.

This brings us to the question of the actual geographical context, or the spatiality of superdiversity in urban spaces. Superdiversity has been discussed mostly in large metropolitan cities or even arrival cities of different groups and types of migrants, such as London or New York. In these cities, there is increased mixing or conviviality of migrant groups (cf. Vertovec, 2015; Wessendorf & Phillimore, 2019). However, although migration is a global matter and the link to the local urban context appears so obvious, a clearer conceptualization of the scalarity of urban diversity, that is, the inter-scalar connection between the local urban transformation and the global phenomenon of migration-led diversity in society has only found little attention with regard to the spatiality of superdiversity. As Hall and Savage point out (2016), the urban theories at the global scale, which discuss issues of global planetary urbanization (Brenner, 2018) or the interconnectedness in the global cities network (Taylor & Derudder, 2015; GaWC), tend to have a stronger economic and economic geographical perspective. As the literature review in Chap. 2 had pointed out, the debates on the space-making or global-city making remain on the global level of either global players and further economic actors, or even on local level focus on the spatiality of corporate and real estate geographies (cf. Hoyler et al., 2018), neglecting somewhat the human dimensions of these global urban context. In contrast, research on superdiversity has been carried out extensively on the urban level, yet the spatial extent from a socio-spatial perspective that goes beyond the local level and theoretically connects to the global scale has been virtually absent from the debates.

Such is the attempt of this monograph to propose and extend one theoretical approach where a connection between the global context and the local spaces is made through combining the global city theory with the transnational migration and superdiversity research. Bringing these two interconnected fields together, it opens the venue to discuss the relationship between migrants *and* cities. By contextualizing the migrant-led diversification to global but also urban spatial processes, this approach bridges and connects the scales of urban diversity. It calls for further researching such inter-scalarity in the study of the spatiality of superdiversity to more holistically capture the global phenomenon that takes shape in the urban local spaces.

9.3 Multidirectional Diversification Beyond the Dichotomy of Above and Below

Migration had been regarded as an inherent part of the globalization process and also the creation of global cities from the beginning of the emergence of the world city theory (cf. Sanderson et al., 2015). The original theory of Friedmann and Wolff hinted at the importance of labor migration in shaping world cities in its physical, economic and social dimensions (Friedmann, 1986; Friedmann & Wolff, 1982), particularly with regard to the dichotomy of "transnational elites" and the "permanent underclass" (ibid: 322) in these cities. Sassen in her seminal work on global cities (1991, reedited 2001) develops the debate further, pointing at the issue of social polarization and the dual migration to global cities. As outlined in the beginning of this monograph, although this nexus between migration and cities is of core interest to migration as well as urban research, it has been far too neglected in the discourses. In fact, though fundamental in the formulation of the world and global cities theory, this particular aspect appears to be long forgotten. Research on the global cities side, the discourse has primarily focused on the economic geographical context of corporate geographies and the interconnectivity of cities, with an increasing focus on the quantification of the cities in the global network (cf. Beaverstock et al., 1999; Carroll, 2007; Derudder et al., 2010; Taylor et al., 2012). When it comes to migration-oriented research on global cities, similar economic geographical contexts also dominate, with transnational professionals being researched predominantly in their context as economic actors and in their business networks (Beaverstock, 2004; Carroll & Fennema, 2002; Morgan, 2001). As pointed out before (Yamamura, 2018), researching transnational professionals have been neglected regarding their role as migrants in the urban context and leaving them out when discussing the spatiality of these transnational migrations in the cities.

However, what these initial approaches toward migration in global cities lack in the consideration, also due to the fact that these theories were developed in a different era of global migration, is the multidimensional diversification of the migrant population and their implications for the diversifying cities. As pointed out by Vertovec in the discussion on the concept of superdiversity and indeed also reflected in population statistics, the migrant populations in global cities (of course, relative to each of their own urban history and migration contexts) are diversifying and the awareness of the diversity of individuals[1] has also been more acknowledged than ever before. Moreover, to better contextualize the production of such superdiversified spaces, where the spatial implications of the migrant-led diversity and the diversification of migrant population become most tangible, the multidirectional diversification has be reconsidered also in its multi-scalar embeddedness of the process of space-making. Different actors at different levels are involved in the processes of space-*making* and space-*shaping*. While the *making* conceptually refers to the processes that contribute to the overall conditions in which

[1] Societal diversity, particularly the discourse on diversity and inclusion, has become ubiquitous, with issues from sexual orientation and gender identities, over different types of disabilities, including neurodiversity, visible and invisible disabilities, to religious diversities and beliefs. However, the growing interest and acknowledgment of these diversity dimensions should not be mistaken with a novel emergence of diversity in society itself as diversities in individuals have been existent throughout the human history. It must not be presumed as a diversification of humans per se but rather the increase in awareness and discourses on diversity dimensions, and through that, the increase of the visibility of diversity in society, respectively the emergence of diversity in societal realities and discourses.

superdiversity occur, such as in the economic geographical contexts of global city making (Hoyler et al., 2018; Yamamura, 2018), the *shaping* of the superdiversified spaces concerns more the direct involvement and practical implications of shaping the urban spaces through spatialized actions and interactions of different actors, respectively transnational practices (Yamamura, 2022c). Space-making and space-shaping processes are indeed not only economic phenomena, and further cannot be decontextualized from global societal changes of the age of mobility and digitalization as well as migration policy issues, which allow or disallow international migration of different groups. Before going into the debates of the structures and agencies of urban superdiversification, this section briefly reviews the findings of this monograph from the perspective of multidirectional diversification.

... From Above

Recent research on planetary or transnational gentrification (Hayes & Zaban, 2020; Lees et al., 2016; Sigler & Wachsmuth, 2016), which is at the nexus of transnationalism and urban studies, points out the spatial expressions of diversity in urban spaces with transnational migrants playing a key role in closing global rent gaps and contributing to the transformation of urban spaces locally. While works on gated communities (Atkinson & Blandy, 2013; Blakely & Snyder, 1997) or the concept of upper-class citadels (Marcuse, 2000) are rather classical approaches to describe physically exclusive spaces, respectively architecturally distinct places within cities for upper-class residents. The upper-class residents indeed also encompass transnational migrants from above, who correspond to and partly mingle with the local peers. Yet, the urban diversification induced by such monetary and economic elite, that is, "from above," brings socio-spatial implications beyond the visible.

As illustrated extensively, such spatial diversity from above are found in areas in the vicinity of embassies and consulates, but also cultural institutions and leisure clubs, well connected to international schools. Through the availability of English-language signs and menus at the shopfront and in restaurants, as well as multilingual services, the overall neighborhood becomes more transnationalized, not only in its physical appearance but also in the socio-spatial practices bound to it. In fact, what has been also analyzed as part of transnational gentrification, the urban landscape becomes an expression of the socio-spatial superdiversity in these cities. These are the places where a diversified group of transnational migrants from above meet and interact, thus, share their socio-spatial everyday practices.

The interesting aspect of the transnationality of this socio-spatial diversification from above is their group constellation that goes beyond the ethno-centric groupist socialization, which is otherwise well documented in transnational migration research. The transnationals from above are predominantly unified in their affiliation to a specific industry and/or to the corporate society, and function beyond ethnic, religious or national affiliations. The corporate affiliation and therewith the privilege of the corporate elite in accessing migration schemes and legal statuses is an important aspect of the class-based transnational migration patterns, and make another specific aspect of superdiversification at the transnationalism from above.

Moreover, the novelty of this present study is the distinction within the often homogenously categorized and characterized group of transnational elites, which illustrate the complexity of spatial superdiversification. While we have the gaijin ghetto migrants who more or less stick to their cocooned comfort zone in such globalized spaces, who reproduce the class-based structures and practices, the Pro-Tokyoites or Pro-Global-Cityites venture out of these areas. These Pro-Global-Cityites keep their more localizing socio-spatial behavioral patterns also beyond one city but rather have embraced it as a lifestyle also in other cities where they have been dispatched. By such socio-spatial behavior they contribute to the interaction and mingling with local populations and local spaces. Commonly, this group of transnational corporate migrants actively and consciously endeavor the connection to local society and culture. Yet, the important aspect here is these socio-spatial activities are still within the realm of their local peers. These spaces thus correspond to those of the local upper class and are not necessarily shared with lower-class locals or other migrants.

These two ideal-typical socio-spatial patterns illustrate that there is a multidirectional socio-spatial superdiversification between a homogenization of elite spaces increasingly observed in many of the global cities, rather becoming universal, and a pro-Global-Cityite transnationalized space in each of the global cities, characterized by a more "local flavored" area connected to the global culture. With the diversification of transnational professionals' group in themselves and the increased awareness and interest in local culture by these multiple migrants, there is a multidirectional superdiversification occurring in these cities even when only considering the dynamics from above.

... From Below

The transnationalism and with it the socio-spatial superdiversification from below are well researched and also more commonly intuitively associated with migrant diversity and cities. Ethnic towns as China Towns, or other ethnic neighborhood like Little Italy, Korean or Indian Towns in many different cities of the world, are prime examples for such transnationalism from below. Often touristified nowadays, but originally developed as labor migrants' areas in cities, they are characterized by the high density of ethnic-focal shops and visible ethnic cultures. Streets are usually full with signs, flyers and posters in the respective language, ethnic themed material culture densely visible on the streets and at the specific shop fronts which is different from the local cityscape. When it comes to actual location within the cities, these ethnic towns tend to be located in less privileged areas, even often coupled with a disadvantageous geomorphological location as these places where areas provided to the urban underclass.

Socio-spatially, so regarding social but also cultural practices in space, these urban spaces are strongly linked to specific ethnic groups or regions. Cultural events as Diwali festivals or Chinese New Year's parade take place, and socio-cultural activities are offered by the respective ethnic or religious groups. Ethnic grocery shops catering traditional food and goods is typical, so are also smaller ethnic minority and migrant entrepreneurial businesses with services focused on the specific needs of their ethnic niche. Travel agents, hairdressers or even special bakeries organizing also catering services for typical weddings and other celebrations are often found in such superdiverse areas and streets (Hall, 2015).

However, beyond these more conventional understanding of ethnic towns and what is visible at the first glance, there is also an increasing superdiversification more subtly occurring in these city areas. Recent empirical studies on the entrepreneurial superdiversity in Glasgow have shown how business-related diversification strategies of ethnic minority and migrant entrepreneurs contribute to the broadening of the niche markets for the entrepreneur. A Polish delicatessen shop could start expanding its products to the Eastern European community beyond the Polish, so can a Chinese shop also be servicing other East Asian and Southeast Asian customers (Yamamura & Lassalle, 2020). Furthermore, middlemen who offer ethnic and ethnicized products to the mainstream market also indicate how the narrower idea of an ethnic business is changing in contemporary societies, contributing to the migrant-led diversification of the local society. With the shops and businesses changing, entrepreneurial superdiversity leads to socio-spatial diversifications of the cities in which the different social and ethnic groups concentrate. As can be seen in the previously mentioned studies on entrepreneurial superdiversity in Glasgow, but also intersectional entrepreneurship in Amsterdam and Hamburg, the spatial expression of superdiversity can be found differently within the same city and even in the same neighborhoods (Yamamura & Lassalle, 2020), calling for a multi-scalar perspective to the multidirectional diversification in cities (Yamamura et al., 2022).

... From the Middle and Beyond

Beyond the dichotomy of the transnationalism and thus socio-spatial diversification from above and beyond, the superdiversity also applies to all the spectra and facets in-between. Middling transnationals, that is, skilled middle-class migrants with decent but not extremely high income, such as language teachers (Conradson & Latham, 2005; Yang, 2022) or lower- and middle-management corporate migrants, do not or cannot necessarily cocoon as the transnational gaijin ghetto migrants. Compared to transnational migrants from above, who are clearly the target of high-end real estates, financial capitals are more limited with these middling migrants. This also results in different consumption and, by that, different socio-spatial patterns that are also specific to their social group. Whereas transnational migrants from above may be capable of dining at high-end restaurants and socializing in expatriate bars, middling migrants would need to choose other locations, especially on the long run on a daily basis. Depending on the still available means, middling migrants may also find international schools too expensive as well as the membership of elite clubs. This then leads to not only socio-spatial practices becoming more localized or detached from gaijin ghettos and other upper-class areas, but also to a more localized socialization, for example, through schools or new encounters in more local spaces, which in turn also has a socio-spatial diversification in the middling group as a consequence (see the case of the demoted transnational professional in Chap. 7). With the growing research on this middling migration, there are avenues of research that need to be further explored. As the accounts of middling and lower-skilled Filipino migrants in Tokyo show, there are further potential for research especially with regard to the consequences of such class- and skill-based transnational migrations on the spatiality of superdiversity.

9.4 Dynamics of Superdiversification of Socio-Spaces: A Multi-Scalar Perspective

Complexity of diversity and diversification has been increasingly pointed out in the research of superdiversity. However, precision and concretization of what exactly the mechanisms and contexts are, in which phenomena such as the above-discussed multidirectional diversifications are embedded, are still lacking. The intriguing perspective on the more ethnographic and social geographical approach to multidirectional diversification in urban spaces is in fact the link to the larger contexts, the inter-scalar nexus between the urban local level and the global migration level. This is where the multi-scalar embeddedness is fruitful and highly relevant (Yamamura & Lassalle, 2022). As much as the local urban transformations have their own dynamics, they are each also embedded in different policies and contexts at different levels.

The multi-scalar embeddedness for transnational migrations is multi-fold and illustrates well how the diversification is influenced and contextualized at different levels:

On the *global level*, as much discussed above especially in the context of the global city research as well as the time geographical constraints of the corporate world for transnational migrations, the global economy as well as economic policy structure defines which cities become global cities and how much they are connected with each other (Brenner, 2018; Sassen, 2002; Taylor et al., 2012). Though decisions are taken on a meso-level of corporations, corporate strategies are part of these global structures, deciding which transnational professionals are dispatched from which location to another. This not only applies to transnational migration above but also for those below, who serve the upper class as well as mid-management and other corporate middling migrants.

Such skill-based global mobility patterns go hand in hand with policies at *the national level*. Designed at the national level, especially migration policies enable or even actively promote migration of highly skilled and also highly affluent migrants to specific countries. Simultaneously the national policies can also allow the migration of lower-skilled workers and/or also restrict them temporarily and spatially, contributing to the dynamics of global mobility (see Yamamura, 2022b, 2022c). To a lesser extent, but still as crucial as for these migrants are policies and legal schemes that enable and constrain middling migrants to move from one country to another (Yamamura, 2022b; Ye, 2016; Yeoh, 2006). Moreover, it is also the national level of economic policies that contribute significantly to an attractive fiscal or legal framework for transnational corporation to enter the domestic market or real estate developers to invest and bring in projects.

At the *regional level*, specific regional policies sometimes also aim at bringing in specific group of people or support specific activities. Recent policy changes toward migrant entrepreneurs in Japan (Yamamura, 2023) are such example. Attracting entrepreneurs specifically in those economic sectors that are in focus for the regional development, prefectures support entrepreneurial economic endeavors of migrants and, more on the social side and in cooperation with the municipalities, the trans-cultural integration and interactions of different groups in the region. The economic and social practices that result from these policies are located in specific spaces locally, and contribute to the expression of superdiversity in the urban or local space.

At the *local level*, the accounts on the difficulties of Pro-Tokyoites or the facilitated integration into specific areas of the city of the gaijin ghetto show how decisions and strategies of further economic and political actors contribute to the socio-spatial patterns. Policy-makers, investors or other decision-makers in urban planning or municipal policies, as well as civil society actors, thus impact the shape of superdiversity in urban space. Intermediary actors of the corporate migration industry, such as relocation and real estate agencies, but also more conventional migration industries considerably contribute to the socio-spatial patterns. They can direct them into specific locations of their assumed preferences and navigating them into upper-class gaijin ghettos despite in some cases desires to venture out or keep them in more remote locations within the global city-region (Yamamura, 2022a). Simultaneously, interactions between the Pro-Global-Cityite, middling migrants and the local population contribute to an additional subtler change in the urban landscape as well as to more diversity when they encounter, an effect on the spatial diversity by conviviality.

Complexity of Spatial Superdiversification

These complex dynamics of multi-scalar structures that make and shape the mobility and social-spatial patterns of the urban space are further enhanced by the diversification of the transnational migrants themselves. As depicted in the course of this monograph, dynamics of socio-spatial patterns in transnational migrants' lives change over the time, along with their career stages, but also importantly in their private lives. Multiethnic, multi-racial, multi-national marriages as well as different types of partnerships have become more common nowadays, and apply to transnational migrants, too. Inter-group marriages or even friendships lead to diversifications of their socio-spatial patterns. Changes in family socialization through children or spouses substantially contribute to differing socio-spatial patterns within the same city. These agencies that transnational migrants still have as individuals with decisions and preferences constitute the counterpart to all the multi-scalar structures that enable and constrain their patterns. Such is how the dynamics of structure and agency create an everlasting dynamical phenomenon of socio-spatial diversification. With the growing generation of migrants who become multiply mobile over their life courses (Zufferey et al., 2021), the diversification of the migrant population over time is inevitable. Adding to the dynamics, further generations of transnational migrants growing up in these transnational spaces—such as the third culture kids—the multidimensional superdiversification of transnational spaces, in particular spanning the increasing number of globally connected cities, will progress even more.

Indeed, spatial superdiversity must be understood as a dynamic process of socio-spatial patterns of a diversifying society which transforms urban spaces differently rather than any kind of end-product of superdiversifying societies. Spatial diversities in global cities are illustrating how spatial superdiversity must be perceived and discussed as a spatial superdiversification. The socio-spatially different patterns and the consequential spatial diversification are inherently connected to the different contexts of differential inclusions and mechanisms of multi-scalar structures and agencies. It is dynamic in its nature, shifting spatially and with time, whereas the socio-spatial dynamics are processes that are influenced by but also find their spatial expressions at multiple levels.

Dynamics of socio-spatial diversification are in fact crucially a temporal-spatial phenomenon as the dynamic elements are that of constant change, activity and progress—which are all embedded spatially and temporally. As the excursion into time geographical perspectives and analysis of transnational professionals' socio-spatial patterns have demonstrated (Chap. 6), the availability, management and allocation of time are crucial in the context of mobilities, bet it within the urban space or as migration crossing national borders. Time has become capital nowadays, which can enable but also constrain individuals in their agency. Such theoretical understanding of socio-spatial diversification processes is also inherently connected to the issue of structuration—the balance or imbalance of structures and agency.

9.5 The Neglected "Other Side of the Coin" of Spatial Superdiversity

At the same time, such dynamics of socio-spatial diversification as discussed above also bring new perspective of hierarchizations of spaces in these global cities. The hierarchies go beyond ethnicity, but along the line of institutional-political embeddedness as migration policies coupled with economic policy give a structural framework to the different migration schemes, enabling or constraining mobility patterns at different levels. Such approaches have recently been taken up by scholars on the migration industries of the highly skilled migration (Cranston, 2016; Koh & Wissink, 2018). The access to global but also local mobilities is bound to the privileged legal status of categories of migrants, especially of the corporate professionals. Moreover, the corporate hierarchy on positions and attached privileges, especially capitals in financial terms, but also in terms of social capital to access local resources becomes crucial for the forms of differential inclusions in global cities where different transnational migrations come together.

Indeed, the global and local contexts and spatiality of superdiversification demonstrate the inherent connection between these socio-spatial dynamics and socio-economic, respectively class- and skill-based behavioral patterns of transnational migrations. It should not be ignored that the changing dynamics in urban spaces, where parts of the cities are transnationally gentrified or in any way transnationalized either by gaijin ghetto or Pro-Global-Cityites, are still processes of the upper class, the dominating financial elite. This inherently means that there are less privileged or affluent local populations are driven out of these areas, and, as the case of the middling and transnational migration from below has shown, less privileged migrant populations are also dislocated to outer areas of the metropolitan region. Such power dynamics are reflected at multiple scalars from the global over the regional to the local spaces, procreating the unequal access to space and reproducing social inequalities. Moreover, as much as the space is an inherently social-spatial issue in human geography (Massey, 2013), the uneven distribution of power over space and mobility is also reflected in the power over time as depicted in the chapter on transitionists, but also in the debates regarding the increasing digitalization of society (Yamamura, 2022a).

Empirical research breaking the online-offline divide in space in the context of superdiversity has emerged (Blommaert & Maly, 2019). However, resonating with the general critique on the discourses on superdiversity that it lacks the critical perspective on the other side of the coin of superdiversity, the access or the lack thereof to the dynamics of spatial superdiversification needs to be paid more attention. After all, as discourses on digital geographies illustrate, the spatiality of superdiversity is inherently connected to the unequal access also to the digital space and digital means. Societal inequalities of migrants are inherently connected to the discourses on the digital divide (Castells, 1996; Dodge & Kitchin, 2002; Graham, 2011) and the im-/mobility dichotomy (Sheller & Urry, 2016; Turner, 2007; Urry, 2007). The development and proliferation of digital technologies and mobility infrastructures have strong implications for the migrants' transnationalism but also for the socio-spatial diversification in cities. Digital means enable and accelerate the transnational connections of cities and impact the local socio-spatial patterns of transnational migrants and non-migrants alike. It is through social media and digital devices that migrants can venture out to unknown lands and explore new grounds, but also keep them shut out of the local happenings and focus on their locally detached communities when preferred. Simultaneously, those who are barred from these means face limitation to their agency, being disconnected and excluded from specific spaces. Through such socio-spatial contextualization and focus on the spatiality, the superdiversity research could and should also contribute to debating issues of inequalities and intersectionalities emplaced in these cities.

From Intersecting Spaces Toward Intersectional Spaces

Diversity comes with issues of intersectionality by default, likewise does spatial diversification come with intersecting spaces. As much as socio-spatial dynamics are bound to power issues, spaces and places are results of continuous struggles over them by different actors contextualized in multi-scalar embeddedness of structures. This goes beyond issues of migration or migranticization of individuals (Dahinden, 2016), which is part of the power struggles in society. Indeed, the debate on superdiversity needs to be de-migranticized in that direction that intersectional issues beyond the migrant background or ethnicity need more consideration. While the reflexive turn in migration studies (Amelina, 2021; Dahinden et al., 2021) is crucial and the social production and reproduction of categories of differences need to be questioned, the purely migration-focused debate on urban diversity and superdiversity needs to be critically reflected, too. Socio-spatial processes of diversity are more complex and issues of inequality and discrimination as relevant as never before in a society in the diversification dynamics.

Especially as transnational spaces are not physically exclusive as in the case of privately owned residential areas (and even them to become privatized exclusive areas preceded power-driven processes of acquisition and occupation), it is a complex and dynamic process of negotiated space-making and space-shaping of different migrant and local social groups. In fact, the negotiation in urban diversity is not necessarily primarily about migrant and non-migrant, it is about what is perceived and taken as the norm of the respective society and those who are not; it is also about who is desirable and visible in urban space and those who are not.

In contrast to simply focusing on the presumed top-down power relation in which dominant capitalist and power elite confines and constraints the powerless of the migrant and urban underclass, there is more subtle differentiation of the involved multi-scalar structures, actors and agents needed. In fact, otherwise one-sidedly underprivileged groups of transnational migrants from below can be intentionally empowered and entitled to maintain their ethnic towns as this adds to a politically and economically desirable effect of diversity in cities. At the same time, middling and less privileged migrants can have more agency through their more localized social networks to access transnational spaces that are barred to the economic elite, who in turn can also be constrained in other ways to venture out of their gaijin ghettos. In any of these complex dynamics of negotiated intersecting spaces, those who are additionally marginalized on the basis of their diversity dimensions, be it gender diversity or sexual orientation, visible or invisible disability, or religious beliefs issues, are disadvantaged in the struggle for the socio-spatial diversification.

It is such intersecting spaces, where different diversities of transnational migrants and social groups come together, that need more attention in the discourses on spatial superdiversity. As noted by Foner et al. in 2019, there is indeed a need to bring in more critical perspectives into issues of intersectionalities. Intersecting spaces can be understood as merely spatial containers or locations in which different migrants and social groups encounter or interact with each other. In contrast to this spatial and locational perspective, what needs more research is the spatially contextualized intersectionality in society, a move from intersecting to intersectional spaces. Intersectional spaces thus are complex socio-spatial constellations, in which power geometries running through our society along different dimensions of social categories of differences intertwine.

Research on spatial superdiversity has sufficiently discussed the positive sides of diversification and the mechanism of a vibrant and beneficial conviviality, such as the ethos of mixing (Wessendorf, 2013). However, what society nowadays needs more is the more critical reflection and analysis of multidimensional frictions and conflicts within urban society in the complex diversification embedded in multi-scalar contexts. While urban social research and particularly critical urban geographers and sociologists have taken up such issues of socio-spatial inequalities, the explanatory mechanism of the inequality appears sometimes too uncomplicated for such complex multi-scalar dynamics of spatial diversification. First attempts have been made in the context of intersectional entrepreneurs and their embeddedness in spatial contexts (Yamamura et al., 2022), discussing the different levels of spatial embeddedness and the multidimensionality of intersectionality as minority entrepreneurs. More need to follow to better grasp, theoretically discuss and eventually also bring betterment to our society in which spatially embedded power relations are creating severe frictions.

References

Amelina, A. (2021). After the reflexive turn in migration studies: Towards the doing migration approach. *Population, Space and Place, 27*(1).

Atkinson, R., & Blandy, S. (Eds.). (2013). Gated communities: International perspectives. Routledge.

Beaverstock, J. V. (2004). 'Managing across borders': Knowledge management and expatriation in professional service legal firms. *Journal of Economic Geography, 4*(2), 157–179.

Beaverstock, J. V., Smith, R. G., & Taylor, P. J. (1999). A roster of world cities. *Cities, 16*(6), 445–458.

Berg, M. L., & Sigona, N. (2016). Ethnography, diversity and urban space. In *Ethnography, diversity and urban space* (pp. 1–13). Routledge.

Blakely, E. J., & Snyder, M. G. (1997). *Fortress America: Gated communities in the United States*. Brookings Institution Press.

Blommaert, J. (2013). *Ethnography, superdiversity and linguistic landscapes: Chronicles of complexity* (Vol. 18). Multilingual Matters.

Blommaert, J., & Maly, I. (2019). *Invisible lines in the online-offline linguistic landscape*. Tilburg Papers in Culture Studies, No. 223. Tilburg University.

Brenner, N. (2018). Debating planetary urbanization: For an engaged pluralism. *Environment and Planning D: Society and Space, 36*(3), 570–590.

Burchardt, M., & Becci, I. (2016). Religion and superdiversity: An introduction. *New Diversities, 18*(1), 1–7.

Carroll, W. K. (2007). Global cities in the global corporate network. *Environment and Planning A, 39*(10), 2297–2323.

Carroll, W. K., & Fennema, M. (2002). Is there a transnational business community? *International Sociology, 17*(3), 393–419.

Castells, M. (1996). *The information age: Economy, society and culture*. Blackwell.

Conradson, D., & Latham, A. (2005). Transnational urbanism: Attending to everyday practices and mobilities. *Journal of Ethnic and Migration Studies, 31*(2), 227–233.

Cranston, S. (2016). Producing migrant encounter: Learning to be a British expatriate in Singapore through the global mobility industry. *Environment and Planning D: Society and Space, 34*(4), 655–671.

Dahinden, J. (2016). A plea for the 'de-migranticization' of research on migration and integration. *Ethnic and Racial Studies, 39*(13), 2207–2225.

Dahinden, J., Fischer, C., & Menet, J. (2021). Knowledge production, reflexivity, and the use of categories in migration studies: Tackling challenges in the field. *Ethnic and Racial Studies, 44*(4), 535–554.

Derudder, B., Timberlake, M., & Witlox, F. (2010). Introduction: Mapping changes in urban systems. *Urban Studies, 47*(9), 1835–1841.

Dodge, M., & Kitchin, R. (2002). *Mapping cyberspace*. Routledge.

Foner, N., Duyvendak, J. W., & Kasinitz, P. (2019). Introduction: Superdiversity in everyday life. *Ethnic and Racial Studies, 42*(1), 1–16. https://doi.org/10.1080/01419870.2017.1406969

Friedmann, J. (1986). The world city hypothesis. *Development and Change, 17*(1), 69–83.

Friedmann, J., & Wolff, G. (1982). World city formation: An agenda for research and action. *International Journal of Urban and Regional Research, 6*(3), 309–344.

Glick-Schiller, N., & Çağlar, A. (2011). *Locating migration: Rescaling cities and migrants*. Cornell University Press.

Graham, M. (2011). Time machines and virtual portals: The spatialities of the digital divide. *Progress in Development Studies, 11*(3), 211–227.

Hall, S. M. (2015). Super-diverse street: A 'trans-ethnography' across migrant localities. *Ethnic and Racial Studies, 38*(1), 22–37.

Hall, S., & Savage, M. (2016). Animating the urban vortex: New sociological urgencies. *International Journal of Urban and Regional Research, 40*(1), 82–95.

Hayes, M., & Zaban, H. (2020). Transnational gentrification: The crossroads of transnational mobility and urban research. *Urban Studies, 57*(15), 3009–3024.

Heil, T. (2020). Comparing conviviality: Living with difference in Casamance and Catalonia. Springer Nature.

Hoyler, M., Parnreiter, C., & Watson, A. (Eds.). (2018). *Global city makers: Economic actors and practices in the world city network*. Edward Elgar Publishing.

Koh, S. Y., & Wissink, B. (2018). Enabling, structuring and creating elite transnational lifestyles: Intermediaries of the super-rich and the elite mobilities industry. *Journal of Ethnic and Migration Studies, 44*(4), 592–609.

Lees, L., Shin, H. B., & López-Morales, E. (2016). *Planetary gentrification*. Polity Press.

Marcuse, P. (2000). Cities in quarters. In G. Bridge & S. Watson (Eds.), *A Companion to the City* (pp. 270–281). Blackwell Publishing.

Massey, D. (2013). *Space, place and gender*. John Wiley & Sons.

Morgan, G. (2001). Transnational communities and business systems. *Global Networks, 1*(2), 113–130.

Sanderson, M. R., Derudder, B., Timberlake, M., & Witlox, F. (2015). Are world cities also world immigrant cities? An international, cross-city analysis of global centrality and immigration. *International Journal of Comparative Sociology, 56*(3–4), 173–197.

Sassen, S. (1991). *The global city: New York, London, Tokyo*. Princeton.

Sassen, S. (2002). *Global networks, linked cities*. Routledge.

Sheller, M., & Urry, J. (2016). Mobilizing the new mobilities paradigm. *Applied Mobilities, 1*(1), 10–25.

Sigler, T., & Wachsmuth, D. (2016). Transnational gentrification: Globalisation and neighbourhood change in Panama's Casco Antiguo. *Urban Studies, 53*(4), 705–722.

Smith, M. P. (2001). *Transnational urbanism: Locating globalization* (Vol. 221). Blackwell.

Taylor, P., & Derudder, B. (2015). *World city network: A global urban analysis*. Routledge.

Taylor, P. J., Ni, P., Derudder, B., Hoyler, M., Huang, J., & Witlox, F. (2012). *Global urban analysis: A survey of cities in globalization*. Routledge.

Turner, B. S. (2007). The enclave society: Towards a sociology of immobility. *European Journal of Social Theory, 10*(2), 287–304.

Urry, J. (2007). *Mobilities*. Polity.

Vertovec, S. (2007). Super-diversity and its implications. *Ethnic and Racial Studies, 30*(6), 1024–1054.

Vertovec, S. (Ed.). (2015). *Diversities old and new: Migration and socio-spatial patterns in New York, Singapore and Johannesburg*. Springer.

Vincent, C., Neal, S., & Iqbal, H. (2017). Encounters with diversity: Children's friendships and parental responses. *Urban Studies, 54*(8), 1974–1989.

Wessendorf, S. (2013). Commonplace diversity and the 'ethos of mixing': Perceptions of difference in a London neighbourhood. *Identities, 20*(4), 407–422.

Wessendorf, S. (2014). 'Being open, but sometimes closed'. Conviviality in a super-diverse London neighbourhood. *European Journal of Cultural Studies, 17*(4), 392–405.

Wessendorf, S., & Phillimore, J. (2019). New migrants' social integration, embedding and emplacement in superdiverse contexts. *Sociology, 53*(1), 123–138.

Wilson, H. F. (2017). On geography and encounter: Bodies, borders, and difference. *Progress in Human Geography, 41*(4), 451–471.

Wise, A., & Velayutham, S. (2014). Conviviality in everyday multiculturalism: Some brief comparisons between Singapore and Sydney. *European Journal of Cultural Studies, 17*(4), 406–430. https://doi.org/10.1177/1367549413510419

Yamamura, S. (2018). The making of transnational urban space: Financial professionals in the global city Tokyo. In M. Hoyler, C. Parnreiter, & A. Watson (Eds.), *Global city makers: Economic actors and practices in the world city network* (pp. 106–123). Edward Elgar Publishing.

Yamamura, S. (2022a). Transnationalism, elite migrants, and the spatiality of superdiversity. In F. Meissner, N. Sigona, & S. Vertovec (Eds.), *The Oxford handbook of superdiversity*. Oxford University Press.

Yamamura, S. (2022b). From global city makers to global city-shapers: Migration industries in the global city networks. *Urban Studies, 59*(11), 2234–2254.

Yamamura, S. (2022c). Transnational migrants and the socio-spatial superdiversification of the global city Tokyo. *Urban Studies, 59*(16), 3382–3403.

Yamamura, S. (2023). The multi-scalar embeddedness of support policies for migrant entrepreneurship in Japan. *International Migration, 61*(2), 67–86.

Yamamura, S., & Lassalle, P. (2020). Approximating entrepreneurial superdiversity: Reconceptualizing the superdiversity debate in ethnic minority entrepreneurship. *Journal of Ethnic and Migration Studies, 46*(11), 2218–2239.

Yamamura, S., & Lassalle, P. (2022). Extending mixed embeddedness to a multi-dimensional concept of transnational entrepreneurship. *Comparative Migration Studies, 10*(1), 1–23.

Yamamura, S., Lassalle, P., & Shaw, E. (2022). Intersecting where? The multi-scalar contextual embeddedness of intersectional entrepreneurs. *Entrepreneurship & Regional Development, 34*(9–10), 828–851.

Yang, P. (2022). Differentiated inclusion, muted diversification: Immigrant teachers' settlement and professional experiences in Singapore as a case of 'middling' migrants' integration. *Journal of Ethnic and Migration Studies, 48*(7), 1711–1728.

Ye, J. (2016). The ambivalence of familiarity: Understanding breathable diversity through fleeting encounters in Singapore's Jurong West. *Area, 48*(1), 77–83.

Yeoh, B. S. (2006). Bifurcated labour: The unequal incorporation of transmigrants in Singapore. *Tijdschrift voor Economische en Sociale Geografie, 97*(1), 26–37.

Yeoh, B. S., & Chang, T. C. (2001). Globalising Singapore: Debating transnational flows in the city. *Urban Studies, 38*(7), 1025–1044.

Yeoh, B., Huang, S., & Willis, K. (2000). Global cities, transnational flows and gender dimensions: The view from Singapore. *Tijdschrift voor Economische en Sociale Geografie, 91*(2), 147–158.

Zufferey, J., Steiner, I., & Ruedin, D. (2021). The many forms of multiple migrations: Evidence from a sequence analysis in Switzerland, 1998 to 2008. *International Migration Review, 55*(1), 254–279. https://doi.org/10.1177/0197918320914239

10

Winding Up

Not concluding the monograph in the sense of closing the debate as it has only just begun but winding it up for the sake of the book and also opening sails to venture out to more to follow, this concluding chapter concisely summarizes the results and presents the direction to which the debate could be further developed in future research. Against the tradition of summarizing according to the order of the results presented (which would repeat the scrutiny from the local to the zooming out to the global), this overview is presented systematically along the lens of zooming in. It begins with the larger picture of the macro-level which has just been discussed in the last previous chapter, encompassing the global spatial superdiversification but then complemented by the overall outlook for national policies impacting the urban space. Zooming back in to the meso-level, the debate on the global city-region and the diversity of actors involved in the spatial diversification is being reviewed. And finally, the micro-level of the space-making by transnational professional within global cities, in particular the case of the global city of Tokyo is then revisited.

Resuming with the penultimate chapter, which is still fresh in the minds, the results of the empirical research on the macro-level suggest

S. Yamamura, *Spatial Diversity in the Global City*, Global Diversities,
https://doi.org/10.1007/978-3-031-64725-3_10

that a global phenomenon of spatial superdiversification is occurring, which in itself is already showing the much-discussed idea of "diversity in diversity" of the superdiversity concept. Two dimensions can be distinguished here, with one being the universal convergence of gaijin ghettos, which can be colloquially termed as the dispersion of expats' ghettos in global cities. These transnational spaces are similar to each other, providing businesses and services universal to the global cities, with leisure activities surrounding corporate and governmental institutions and depicting a Western way of life abroad. Maybe less fascinating or surprising in its existence at first sight, but the in-depth research has shed light on the complexity of contexts in which this peculiar ghettoization is actually embedded. The socio-spatial patterns of some transnational migrants leading to the gaijin ghetto are not a simple top-down or endogenous choice of the "transnationalism from above," but the individuals in them are constrained or at least contextualized in an environment which also exogenously contribute to their spatial exclusion, that is, especially the overall corporate strategies of short- and mid-term dispatchment of migrants in the global economy. The other dimension of spatial diversification is the universality of superdiversification of Pro-Tokyoite transnational spaces both within each global city and also within the network of global cities. With the tendencies of Pro-Tokyoites to socio-spatially localize and mingle more with the local society (yet within the same socio-economic class), and doing so in each of the global cities they migrate to, an extreme variety of mixtures of Pro-Global-Cityites spaces are being created throughout the network of global cities. Such Pro-Toykoite socio-spatial pattern showcases the spatial superdiversification of global cities in times of transnational migrations.

These mixtures of global universality and yet local diversifications are constantly moving as social dynamics and thus socio-spatial patterns are not static. The transitionists' case has demonstrated the dynamic, but also macro-level contexts are major parts of the dynamism. As much as transnational professional migrants have migrated away from Tokyo in its post-bubble period, there are already hints of transnational corporations, and along with them transnational professionals, also migrating from London due to the impending Brexit. Simultaneously, Tokyo has again gained some (short-lived) momentum with Tokyo 2020, the mega-event of the Olympic Games in 2020 postponed to 2021. On the one hand, it was backed up by active economic policies of establishing special economic zones to attract foreign investment and to encourage transnational corporations to return. On the other hand, it was accompanied by a placative policy turn to diversity using the sub-motto of "unity and diversity" which enfolded in a rather unintended negative headlines and conflicts on gender and diversity issues. Such dynamics of the national policies and the global economy whatsoever contribute to the social and spatial diversification in the global cities.

Concerning the meso-level, it should be noted that although the flows of money and accordingly people follow the global economic dynamics, regional and local economic actors remain drivers in making spaces in and for the global economy, such as providing corporate real estate and subsidies in urban and regional development to attract global players. The presented research has shown that the global city-regions are part of the global dynamics and are crucial specifically for the socio-spatial patterns of transnational migrants. Going not only beyond national borders but also beyond administrative urban borders, transnational migrants contribute to the spreading of the transnational space from the global city center to the surrounding region. At the same time, regional and urban developers create new spaces with their large-scale projects, attracting transnational migrants to migrate to them. They can be regarded part of the actors contributing to the global Roppongization, attracting foreign investments and by that also highly skilled migrants into these regions. The global city-region remains the main arena of interplay between macro-level political and economic dynamics and the micro-level socio-spatial behavior and decisions of the actors involved.

As a last step, zooming into the micro-level, that is on the side of the urban space in which transnational migrants and also further individual actors are embedded, the presented research has demonstrated that more insight into the nature of these spatially bound social diversifications is needed. As much as global cities have been criticized for producing social polarizations, especially of the migrant population, there is an even stronger requirement to reconsider the individual needs, desires and also structures in which the allegedly capitalist transnationals from above are embedded themselves. The urban space has become a highly politicized context of urban development, carrying issues of deprivation and dispossession of social group. There is certainly a need for more in-depth and objective empirical research into the transnationalism "in-between" on the individual level, shedding light on the mechanism creating socio-spatial patterns. In this context, the interactions of transnational spaces of different migrants and urban actors, who contribute to the actual spreading of "diversity in diversity," especially need to be more closely studied. To make the best of social and spatial superdiversification, the complexity of transnational migration should not be left as the backside of the coin, but be more systematically looked into to find strategies and policies to embrace the benefits of the diversification.

Finally, Tokyo is certainly unique in its urban history and as part of the global economic dynamics as a forerunner of the "Asian flying geese," it is also the first to somewhat have lost its status and is currently still struggling to keep up with the dynamics. However, in the end, as it may not remain the first and the only one to lose its status as the prime global city, the case of Tokyo could deliver valuable empirical insights for future reference. The study could become a good beginning to look into the dynamics of the socio-spatial diversification of global cities and the changing constellations of the network of global cities. Especially with the emergence of non-postcolonial cities becoming global cities, the understanding of societal and spatial diversification could contribute to better accommodating the increasing complexity of transnational migrations in non-English-speaking local societies.

Bridging several gaps in transnational migration research, the presented research may have not fully closed these gaps, but at least has surely contributed to identifying these conceptual and empirical vacuums, and also delivered unique and valuable empirical insights into the spatial diversification by studying a long neglected transnational migrants group. In times of superdiversification of society, this work calls for a novel approach into the research field of transnational migration. Social diversification goes hand in hand with spatial diversification and vice versa. The methodological and conceptual approach applied in this work shows an option to better break down and, thus, better comprehend the complexity of the dynamic socio-spatial superdiversification in the increasingly interconnected world of global cities.

*   *   *

As mentioned in the Preface, the main fieldwork for this monograph was conducted before the COVID-19 pandemic, where socio-spatial mobility and social practices of urban diversity had been starkly different. As discussed before, the paralyzing shock of deprivation of freedom to mobility and interaction was universal. Yet, observing the very minority of the highly privileged transnational professionals has also shown how such restrictions still impaired their practices but also affected them less than other social groups. Jetsetters were still somewhat jet-setting, finding their own new dynamics of mobility and also being financially capable to accommodate the new work conditions. Additionally, while tourists and average travelers were barred from entering countries, the first cohort of privileged entry during the easing phase of border controls was addressing economically important business persons.

While the interim phase directly after the lockdown phase of the pandemic brought up serious concerns regarding city life with the many closings of small retail shops and gastronomy during the pandemic, the urban life nowadays in the post-pandemic time seemed to have caught up again. In fact, there is a rather shocking turn back to the normal as before the pandemic, with decreasing number of individuals wearing masks, keeping physical distances or even avoiding social events. Yet, some aspects of society have certainly and irrevocably changed—not necessarily because of the pandemic, but rather due to all other political, economic and social factors which have emerged and are ongoing in our global society. Issues of racial and intersectional discrimination, far-right movements and issues of radicalization and extremism, creating a diversity- and migration-hostile environments in times of refugee and migration, but also the Ukraine war, the Middle East conflict and economic crises contribute to a different landscape than pre-2020.

The question how all these impact the presented results and the discussions of this monograph is a fairly difficult one to answer. However, what can be said for sure is that the overall trends in the universal conversion of spaces within the network of global cities on the one hand and the local diversifications on the other remain evident. Expat gaijin ghettoes have been even strengthened during the pandemic, developing its own ecosystem of information and supports; while those who are keen to venture out have the means to do so more than ever. In fact, the spread of social media and the ubiquitous usage of digital devices contribute to the twofold socio-spatial patterns as each of the preferences are catered and patterns even more so perpetuated. Social media, as they are nowadays, do not only radicalize social groups in their political views, being fed similar information or misinformation and connected to communities of similarly opinioned people. They also contribute to the perpetuation of also expat communities' specific gaijin ghetto patterns, but also can become opportunities to connect beyond one's comfort zone with local peers—depending on the individual's preference and affinity.

Details of the actual urban areas within Tokyo or within the overall network might and will be changing—there are shifts in the socio-spatial patterns for sure. For example, due to simple factors, such as the weak currency (the Japanese Yen), those who are paid by a stronger foreign currency or under the economic stability of their employers, which tends to be the case with migrant governmental officials or intra-firm transferees, the capital and with it the opportunity rises to venture out into spaces that were more exclusive to only the highly privileged transnational professionals. Such change can shift the boundaries of gaijin ghettos as well as Pro-Tokyoite spaces; not only to each other, but also to the spaces of middling transnational migrants. Similar changes will be occurring with the development of new desirable urban hotspots created by urban developers or places orchestrated by mega-events or other large-scale projects. Such changes are natural and therefore predictable or inherent part of the socio-spatial diversification of global cities. Same goes to the global dynamics of the attractiveness of global cities depending on the political, economic or even environmental changes. Examples for such short- and long-term changes are the sudden exodus of transnational professionals out of Tokyo during the Fukushima crisis or the more sustainable dwindling away of expats out of Hong Kong since the failed Umbrella Revolution.

Such empirical results from a previous fieldwork might not be sustained for a long time as times change quickly as we can see in our everyday coping with multiple crises that have just arisen within less than one year. The main message of this monograph on the spatial diversification trends and the mechanism to the space-making of transnational migrants, however, along with the theoretical contributions on the multi-scalar embeddedness and the temporal-spatial contextualization necessary for understanding the spatiality of superdiversity will be valid as before the pandemic. The call for a different and more concrete approach to investigating urban super-/diversity as well as the urge to focus future research surrounding superdiversity more on intersectionality and intersectional spaces will probably and unfortunately stay as timely and relevant in the next decades. As a matter of fact, it would be a desirable shift and hope to see society and academia taking this monograph as a relic of the past and the calls to become obsolete.

Index[1]

[1] Note: Page numbers followed by 'n' refer to notes.

S. Yamamura, *Spatial Diversity in the Global City*, Global Diversities,
https://doi.org/10.1007/978-3-031-64725-3

SPRINGER NATURE

GPSR Compliance

The European Union's (EU) General Product Safety Regulation (GPSR) is a set of rules that requires consumer products to be safe and our obligations to ensure this.

If you have any concerns about our products, you can contact us on ProductSafety@springernature.com

In case Publisher is established outside the EU, the EU authorized representative is:

Springer Nature Customer Service Center GmbH
Europaplatz 3
69115 Heidelberg, Germany

The manufacturer's authorised representative in the EU is Springer Nature Customer Service Centre GmbH, Europaplatz 3, 69115 Heidelberg, Germany. If you have any concerns regarding our products, please contact ProductSafety@springernature.com

Printed and bound by CPI Group (UK) Ltd, Croydon, CR0 4YY
01/10/2026
02225954-0002